HOLLYWOOD GAMERS

ROBERT ALAN BROOKEY

HOLLYWOOD GAMERS

DIGITAL CONVERGENCE IN THE FILM AND VIDEO GAME INDUSTRIES

INDIANA UNIVERSITY PRESS
Bloomington and Indianapolis

This book is a publication of

Indiana University Press
601 North Morton Street
Bloomington, Indiana 47404-3797 USA

www.iupress.indiana.edu

Telephone orders	800-842-6796
Fax orders	812-855-7931
Orders by e-mail	iuporder@indiana.edu

∞ The paper used in this publication meets the minimum requirements of the American National Standard for Information Sciences—Permanence of Paper for Printed Library Materials, ANSI Z39.48-1992.

Manufactured in the United States of America

LIBRARY OF CONGRESS CATALOGING-IN-PUBLICATION DATA

Brookey, Robert Alan, [date]
Hollywood gamers : digital convergence in the film and video game industries / Robert Alan Brookey.
p. cm.
Includes bibliographical references and index.
ISBN 978-0-253-35524-9 (cl : alk. paper) — ISBN 978-0-253-22231-2 (pb : alk. paper)
1. Motion pictures and video games. 2. Motion picture industry—Technological innovations. 3. Video games industry—Technological innovations. 4. Convergence (Telecommunication) I. Title.
PN1995.9.V46B76 2010
338.4'77948—dc22 2010002058

1 2 3 4 5 15 14 13 12 11 10

To Gary Schwartz,

for handing me the control pod

to his PlayStation 2

and asking,

"You want to try?"

CONTENTS

ACKNOWLEDGMENTS

This project began as an independent study that I conducted with one of my former graduate students, Paul Booth. The study proved to be very productive, and we collaborated on an article that appeared in *Games and Culture* (the second chapter of this book is my expansion of our original analysis).[1] Paul's contribution, however, is not limited to this chapter. It was his enthusiasm and interest in our independent study that convinced me that I should pursue the convergence of the film and video game industries further.

Many of my undergraduate students also contributed to the preparation of this study as part of their practicum requirement. Unfortunately, my institution has recently switched to a new software program for academic record keeping, and I was unable to retrieve the names of all the students who contributed. To name a few and miss the rest would give the false impression that I did not value the help of those I failed to mention, so I will simply thank them collectively for their assistance. The interactions that I had with these students were invaluable: not only did they help me to turn the games into analyzable texts, but they also provided me with great insights into the practices of gaming.

The book manuscript was completed while I was on sabbatical from Northern Illinois University, and my colleagues at this institution also supported my efforts. While I was working on this project, David Gunkel was studying video games as well, and he proved to be a valuable resource. I thank my other colleagues for their support; they know who they are. I also thank Janice Hamlet for the very, very special pen.

Working on a book can yield some unexpected and welcome results. This book gave me the chance to reconnect with an old friend, Diane Miller. Diane and I had become close friends in graduate school, but had drifted apart. She and I had collaborated in the past, and she was always able to work wonders on my pitiful prose. I knew that she was doing some freelance editing, and so I decided to have her work on this book. Diane helped me in ways I could not have imagined. The conversations that we had in the editing process spilled over into other aspects of our lives, and I was reminded of why we were once close friends. I offer my most heartfelt thanks to Diane.

HOLLYWOOD GAMERS

ONE

PLAYING TOGETHER

The film *300* (2007) surprised everyone. Its subject—the ancient battle at Thermopylae—was hardly blockbuster material. After all, the sword-and-sandal genre had not enjoyed much popularity since Jason and his Argonauts battled Ray Harryhausen's stop-motion animated monsters. Granted, *Gladiator* (2000) did quite well at the box office, but it had Russell Crowe; *300* had Gerard Butler, not an A-list actor at the time and certainly not a box office draw. Still, the film drew an audience. It opened on March 11, 2007, and pulled in $70 million over the weekend, then went on to generate over $200 million in U.S. domestic box office receipts, exceeding all expectations. It more than doubled those receipts in the foreign markets and closed its theatrical run only after racking up $456 million at the box office.[1]

This film had executives in Hollywood scratching their heads, trying to comprehend, and hoping to duplicate, *300*'s success. Ron Grover, the entertainment correspondent for *BusinessWeek*, argued that *300* offered several lessons:

> Sometimes it really is about what you put on the screen, and maybe you don't need to put as much up there as you might think. As far as epic wannabes go, *300* is modest, yet audiences are eating it up. The nonstop action came from computers, the actors were, well, wooden, and still the trailers and commercials were mesmerizing. Sometimes a great visual is worth more than heavyweight actors and a legion of writers.[2]

Indeed, most of the action in *300* was filmed in front of a green screen, with computer-generated imagery (CGI) added in post-production—a fact not lost on the film's detractors. Stephanie Zacharek of *Salon* wrote, "*300*, even with its impressive vistas of computer-generated soldiers, is just a throwaway epic."[3] A. O. Scott, a critic for the *New York Times*, was equally unimpressed: "I would happily pay a nickel less, in quarters or arcade tokens, for a vigorous 10-minute session with the video game that *300* aspires to become."[4]

Scott's dismissal, however, may actually speak to *300*'s appeal. Exit polling on the film's opening weekend indicated that the audiences were predominantly male, and half the audience members were below the age of 25. Therefore, the film appealed to the same demographic that serves as the target market for many video games. In addition, *300* did not aspire to be a video game, because it already was one, released about one week prior to the film on Sony's PlayStation Portable (PSP). Indeed, *300* signifies a growing trend: the convergence of the film and video game industries and the growing practice of turning films into video games.

The significance of *300* in relation to this trend was not lost on Clint Hocking, a creative director for the video game company Ubisoft, who wrote in his blog:

> [W]hile convergence is the buzzword that everyone whispers around the boardroom and at the shareholder meetings, the creatives [*sic*] in both Hollywood and the game industry are busy proving why convergence is little more than a buzzword. *300* is—in fact—the movie that proves that convergence between film and video games is impossible.[5]

Hocking goes on to formulate an argument based on points about art, narrative, and aesthetics; he even references Plato for good measure. Unfortunately for Hocking, *300* made a lot of money, and in the entertainment industry money matters. In fact, Thomas Tull, the executive producer of *300*, proved that the convergence between the film and video game industries is more of a reality than Hocking cares to admit. After *300*'s success, Tull closed a multimillion-dollar deal to start a video game company.[6]

In contrast to *300*, DreamWorks SKG never really lived up to industry expectations. When the studio was formed in 1994, it was expected to produce quality content for a variety of media. After all, the studio combined the talents of Steven Spielberg, former Disney executive Jeffrey Katzenberg, and music mogul David Geffen. Although the studio developed a variety of media, including video games, films were its major product. DreamWorks was not without its successes: *Saving Private Ryan* (1998) won an Academy Award for Spielberg, *American Beauty* (1999) won three Academy Awards, and *Shrek* (2001) proved to be very lucrative.

After a decade, however, the partners of DreamWorks SKG decided to sell the film studio to Viacom and go their separate ways. Just before the deal with Viacom was closed, the video game company Electronic Arts (EA Games) announced that it had signed Spielberg to a three-game deal.[7] Under this deal, Spielberg would set up offices in EA's Los Angeles studio and develop three original video games. EA Games would retain the rights to these games, but Spielberg would be able to develop the titles into television shows and films.

This was not Spielberg's first venture into the video game business. In fact, one of the earliest video game spin-offs of a popular film involved Spielberg, though it proved to be a complete disaster. After *ET*'s theatrical release (1982), the executives at Warner Entertainment were anxious to sign Spielberg to a production deal; to sweeten the deal, Warner promised to produce an *ET* video game. Warner had purchased Atari from Nolan Bushnell in 1976, and this was one of Warner's first attempts to use the video game company to negotiate a film deal. In order to deliver the game in time for the Christmas shopping season (one of Spielberg's demands), the game's production was rushed. Consequently, the *ET* game was poorly designed and generally rejected by the children for whom it was marketed.

Atari, banking on the success of the film and expecting a run on the *ET* game, had overproduced cartridges and was stuck with a warehouse full of stock it could not move. To clear the warehouse, the company contracted to have the cartridges buried somewhere in the New Mexico desert.[8] The failure of this particular video game was due to the fact that the experience of the game was far from the experience of the film. *ET* was a film to which many children had a strong emotional attachment, particularly with the alien who seemed to exhibit human characteristics. A video game portraying *ET* as a two-dimensional avatar with limited expression and limited movement did little to elicit such attachment or emotion.

Another early example of the failure to converge the film and video game industries was the *Super Mario Brothers* movie. In contrast to *ET,* the *Super Mario Brothers* film (1993) was an attempt to use video game content to sell a film, rather than the other way around. The film was based on the *Donkey Kong* video game, which was a breakout hit for Nintendo and one of the first video games to incorporate narrative elements.[9] As narratives go, the one in *Donkey Kong* was rather simplistic: the game character Mario had to save his girlfriend, Pauline, from a gorilla. Mario had to climb a series of ladders and levels in order to reach Pauline, while the gorilla hurled barrels and other obstacles in his path. Obviously, the game provided a rather limited narrative arc with which to span a full-length feature film, and consequently, the film was not well received. As Janet Maslin, writing for the *New York Times,* observed:

> This bizarre, special effects–filled movie doesn't have the jaunty hop-and-zap spirit of the Nintendo video game from which it takes—ahem—its inspiration. What it has instead are a weird, jokey science-fiction story, *Batman*-caliber violence and enough computer-generated dinosaurs to get the jump on *Jurassic Park.* Eleven-year-old boys, the ideal viewers for this vigorous live-action comic strip, will no doubt be impressed with the expense and energy that have gone into bringing *Super Mario Brothers* to the screen.[10]

Considering the dismal box office performance (the film generated only $20 million in revenues, while it had an estimated budget of $42 million), it seems the eleven-year-old boys were unimpressed as well.[11]

Given these disasters, it may seem surprising that film studios would still want to do business with the video game industry. Yet, over the years, the technology of video games has improved significantly, allowing games to offer cinematic visuals and complex narratives. In other words, video games have become more like movies, and consequently more movies have become video games. Of the top ten highest-grossing films of 2006, eight had video game releases, and one (*Casino Royale,* 2006) might have had a video game release had the license for the James Bond franchise not been in transition from EA Games to Activision.[12] The video game releases of 2006 were hardly an anomaly: the previous year, seven of the top ten grossing films had video game releases, and the same was true of 2007.[13] Perhaps as the exception that proves the rule, when *The Dark Knight* (2008) was setting box office records, the absence of the video game was noted:

> Why is there no video game based on "The Dark Knight"? For the first time in the film-franchise's history, the caped crusader flew into movie theaters without a video game attached to his utility belt. Despite a plethora of "Dark Knight" action figures, bobble-heads and T-shirts sweeping in Bat-dollars beyond the film's $400 million record-smashing box office, no "Dark Knight" game is following suit. Whatever held things up caused about $100 million in sales to be missed, according to estimates.[14]

Indeed, the release of the video game spin-off has become a common marketing practice, one that carries with it a significant revenue stream.

Granted, not all films are marketed in this way. Independent and foreign films seldom have video game spin-offs; for example, there is no video game for *The Hours* (2002). Yet many action/adventure films and many films marketed to children and families include a video game component. Furthermore, it is often these types of films that generate the most income at the box office; they are referred to as "tent-pole" pictures because their revenue is used to support the rest of the films on a studio's production roster.

In addition, these tent-pole pictures are often parts of franchises: a series of films, including sequels and sometimes prequels, which are then spun off into a variety of different media and licensed products. In 2006, two of the top ten films were sequels of previous films (*Pirates of the Caribbean: Dead Man's Chest* and *X-Men: The Last Stand*), and two others drew on established content provided by the previous James Bond and Superman films. In 2007, four of the top five highest-grossing movies were sequels, and all four had video games.[15]

Video game spin-offs are often tied to the most successful films in the market and are an important tactic in the larger marketing strategy of establishing a film franchise. Indeed, where the most profitable films are concerned, video game releases are becoming the rule rather than the exception. In spite of Hocking's claim to the contrary, convergence has already occurred. Indeed, shortly after he posted his complaint about convergence, Hocking's own company, Ubisoft, signed on to develop the video game for James Cameron's film *Avatar* (2009), the first film Cameron directed since *Titanic* (1997).[16]

This book explores the convergence of the film and video game industries. Specifically, I will examine the practices that turn film content into video game content, and the industrial conditions that inform these practices. While video games are also sometimes turned into feature films, this practice is neither as common nor as lucrative. Although there are exceptions (*Lara Croft: Tomb Raider*, 2001, for example), most films that are developed from video games are not assigned large budgets and do not function as the tent poles for major studios. Therefore, this study will focus on video games that are spun off from films, a practice that has become popular in the twenty-first century. To understand how this convergence has come about, it is important to consider two factors. First, certain technological advancements, specifically the development of the DVD, have facilitated convergence. Second, specific business conditions have allowed the film and video game industries to work cooperatively.

Lairs, Lasers, and the DVD

In 2005, Hollywood studios witnessed the largest drop in box office revenues in over 20 years, with summer attendance particularly low. Summer is the most important time of year for Hollywood studios, because this is when they release most of their tent-pole films. Market research indicated that the target audience for the summer film fare had chosen home entertainment options instead, primarily DVDs and video games.[17]

The box office has always been one of the more unpredictable revenue sources for the film industry, so studios have come to depend on other markets. Home video is the most lucrative of these markets, because films usually make more money on video than they do during their theatrical release. The revenue generated from video game licensing fees, however, is also significant.[18] Although box office receipts rebounded in 2006 and set records in the summer of 2007, studios are still dependent on these alternative sources of revenue.

For several years, the video game and home video markets were distinct in terms of both content and the physical products themselves. Home videos were manufactured, distributed, retailed, and viewed on VHS tapes. Video games, on the other hand, were manufactured, distributed, retailed, and played on ROM cartridges and CD-ROMs.[19] In 1997, however, DVDs were introduced to the consumer market, and at this point the film and video game industries began selling very similar products. The DVD's impact on the home video market is well documented, but it has also had a significant impact on the video game market.[20] Most home video game systems now use the DVD format, and most video games are published in this format. Consequently, once the DVD was adopted by both the home video and the video game industries, important practices in these two industries dovetailed. To put this in perspective, it will be helpful to review briefly the history of video games and home video.

When the video game business began in the 1970s, there were two distinct yet overlapping markets: arcade games and home gaming systems. Arcade games were housed in large cabinets and placed in amusement parlors alongside pinball machines and pool tables. At about the same time, Magnavox began marketing a home gaming system called Odyssey.[21] Many game titles moved from the arcade to the home systems, and companies like Atari released products for both markets. These markets, however, ultimately diverged, and the home gaming market came to dominate the video game industry. As is often the case, the ultimate dominance of the home gaming market was a matter of economics. Software sales produce the most revenue for the video game industry. Gaming consoles are often retailed as loss leaders, and profits are generated from the licensing and sale of software.[22] While the arcades can promote a particular game, they generate revenue at only a dollar or two per play; new game software can start at price points of $40 and above.

For many years, both the home gaming and arcade markets relied on technology that offered only primitive graphics. As the technology improved over the years, so did the image quality of the games. Polygons are the unit of representation in video games; these polygons are programmed as data and stored on software, and the game systems then process the data and render the images on the video screen.[23] As the images become more detailed and the graphics more complex, more polygons are used and more data are needed. Therefore, advancements in video game technology have required the development of better hardware, capable of processing larger amounts of data, and of software capable of storing and delivering these data.

In 1983, it appeared that laserdiscs would significantly improve the quality of video games. In that year, the game *Dragon's Lair* appeared in arcades. *Dragon's Lair* represented a significant advancement in video games because it was based on laserdisc technology that offered full-motion video, rendering complex, cinematic images. *Dragon's Lair* was "cinematic" in the Disney sense of the word—that is, literally. The game used animation produced by Don Bluth, who had previously worked for the Disney studios, where he contributed to the production of *Sleeping Beauty* (1959), *The Rescuers* (1977), and *Pete's Dragon* (1977).

Dragon's Lair was based on Bluth's cel animation, so the game looked like a Disney movie. The game allowed players to interact with a fully animated avatar (a knight named Dirk the Daring) and navigate him through various scenes in order to save Princess Daphne. The interactivity was rather limited and game play could be frustrating: the player had to make the correct choices to move Dirk in the proper manner at strategic points in the game or he would die and the game would be reset. Even with these limitations, however, the game provided players with a new and exciting visual experience, and it became extremely popular.

Generally, a successful arcade game was repackaged for the home video market. As William Hunter notes, there were many versions of *Dragon's Lair* released for home video systems, but most of these proved disappointing.[24] In most cases, the game was released on home gaming systems that did not use laserdisc technology, so these versions did not capture the dynamic imagery that made the arcade game so popular. Unfortunately, the laserdisc technology that made *Dragon's Lair* visually dynamic did not make a successful transition to the home video market.

Laserdisc technology, originally marketed as an alternative to the videocassette recorder (VCR), was known to have performance problems even under the best conditions. Yet video game arcades hardly provided the best conditions, as frustrated players would take out their anger by kicking or hitting the housing cabinets. This type of physical abuse would often cause the laserdisc players in the

Dragon's Lair cabinets to malfunction, and the games were frequently out of order. When a laserdisc home gaming system called the Halcyon was released to the market in 1985, the $2,500 price tag proved to be prohibitive.[25] Again, the laserdisc technology was the problem because the disc drives added to the cost. Although the Halcyon was also promoted as a laserdisc player for home video, this function did not provide much incentive for consumers.

Laserdiscs were never able to capture a significant consumer base, and their adoption was limited to technophiles.[26] This limited consumer base did not motivate Hollywood studios to release many film titles on laserdisc and, in a circular manner, the limited title selection further discouraged consumer adoption. In other words, the laserdisc-based home gaming system failed due to the limited consumer base for this technology in the home video market. It would take another disc format to overcome this spiral of failure.

In contrast to the laserdisc, the digital video disc (DVD) was the first fully digital format to succeed in the broader consumer market. As David Sedman notes, the DVD format was developed through the collaborative efforts of Philips NV, Sony, and Toshiba.[27] Originally these three companies were working on different digital video disc technologies, but they decided not to market competing products, instead forming a coalition to establish a single industry standard that would become the DVD format. The DVD was thus allowed to emerge into a noncompetitive market, unlike the early years of the home video market, when Betamax went head-to-head against VHS, and VCRs battled for market share with laserdisc players.

Sedman argues that DVDs benefited from this coalition standard because it allowed the format to be brought to market more quickly. Indeed, although the format was only introduced to consumers in 1997, by 2001 the Consumer Electronics Association (CEA) reported that DVD players had reached a 21 percent household penetration rate in the United States.[28] By 2006, the CEA reported that 75 percent of consumer households in the United States owned a DVD player.[29]

Clearly, DVD players have been highly successful in the home video market, but the hardware is only part of the story. The format has also had a significant impact on the software market. Part of the appeal of the DVD format is its superiority to VHS tape, but in addition to the improved video and sound quality, DVDs have been able to deliver a lot more than just movies. The DVD format holds a substantial amount of information and can therefore offer features seldom available on VHS tape. For example, DVDs often contain running commentaries by directors, deleted scenes, interviews with actors, and "Easter eggs" (hidden features that first appeared in video games).

The impact of these special features has been duly noted by Elvis Mitchell in the *New York Times:* "The esoterica of film culture, formerly consumed by a moneyed geek elite, is now aimed directly at—and snapped up by—the broader public."[30] Indeed, when Disney's home video distribution arm surveyed consumers, it found that 63 percent considered extra features to be an incentive to purchase DVDs.[31] These features have been used to increase a film's profitability not only by repurposing a film for the home video market, but also by allowing studios to repurpose the same film content multiple times. For example, *The Lord of the Rings* (*LOTR*) film franchise produced only three theatrical releases, but Amazon.com offers multiple DVD products, including individual releases for each film, extended editions, collector's editions, and gift boxes. In addition, DVDs are usually priced in a way that makes consumers more willing to purchase rather than rent these products.

In contrast to VHS tapes, which were marketed to encourage rentals, new DVDs are usually released at a price point that encourages retail sales. As Wilson Rothman notes, with most DVDs priced between $10 and $20 and most new rental prices starting at $4 (excluding late fees), many consumers opt to buy DVDs.[32] Because consumers are choosing to buy movies, their options are no longer limited to Blockbuster, Hollywood Video, or the mom-and-pop rental store. DVDs can be purchased in a variety of retail contexts, including bookstore chains, big-box electronic stores, discount retailers, and even drugstore chains. In addition, consumers can purchase DVDs at Amazon.com and other online stores. Consequently, these additional options have significantly eroded the rental market: Blockbuster posted eight straight years of net losses, and Movie Gallery (which also owned the Hollywood Video chain) had to file for bankruptcy.[33]

In addition to its influence on the home video market, the DVD significantly changed the video game market as well. In 2000, Sony released the PlayStation 2 (PS2), a gaming system that offered faster processor speeds and significantly improved graphic capabilities. As the first video game console to include a DVD drive, the PS2 was marketed as a state-of-the-art gaming system *and* a DVD player. A year later, Microsoft entered the video game market with its Xbox, a console that also included a DVD drive that could play games and home video. In terms of its overall impact, however, it is impossible to ignore the significance of the PS2. With an installed worldwide user base of around 120 million, the PS2 is one of the most successful game systems in the history of the industry.[34] Nintendo proved to be the last console producer to adopt the DVD format, which was incorporated into its highly successful Wii gaming system.[35]

After the DVD gained a foothold in the gaming market and video game software began to appear on DVD, the practices of the video game and home video industries converged in an important way, as both began producing and marketing the same object. Apart from the code burned into the discs, home video DVDs and video games are the same physical product, and the factories that duplicate and produce DVDs can also produce video games.

The manufacturing, packaging, and retailing of home video and video games have thus aligned in an unprecedented manner. Both DVDs and video games are usually packaged in the same slim plastic casing, although special editions of both can feature different types of packaging. In retail stores, DVDs and video games can now be stocked on the same type of shelving, and in big-box stores such as Best Buy, these two products are often shelved in close proximity. In addition, some video game stores have provided a secondary market for used DVDs. These stores actually allow customers to trade in used DVDs for credit toward the purchase of video games; in these stores, DVDs and video games have become exchangeable commodities.

Because of the similar packaging, the close proximity of video games and DVDs in retail spaces also augments the film studios' marketing efforts. For example, studios carefully select the "one-sheet," the visual image that appears on the film's posters, advertisements, and billboards. This image often serves as the cover art when the film is released on DVD, and a similar image is often used for the video game packaging. In the case of the first *Spider-Man* (2002) movie, for example, the poster featured a close-up of Spider-Man's mask, and a similar image appeared on the cover of the video game box. Indeed, for every *Spider-Man* film released, there has been a corresponding video game, and the packaging of each game reproduces visual elements of the posters of the corresponding film.

This use of visual marketing means that the retail space devoted to video games can also function as advertising. For example, the video game for the third installment of *The Lord of the Rings* trilogy, *The Return of the King* (2003), was released about a month prior to the film's release in theaters. During this time, the presence of the video game on store shelves, with the cover art depicting the main characters, served as a reminder both that the film would soon be in theaters and that the video game could be experienced now. This is but one example of how the video game and film industries both function as cultural industries, and as cultural industries, they confront similar business conditions and similar challenges.

Business Conditions, Labor, and Licensing Practices

In addition to the role of technological developments, the similar business practices of the film and video game industries have also facilitated convergence. For example, the costs of film production are entirely up-front. Indeed, millions of dollars can be spent on the development of a film even before the cameras begin to roll.[36] Film producers must secure funding to begin the actual making of the film, most of which is then spent in the processes of production and post-production. Once the movie is finalized, however, it is not that expensive to reproduce and distribute prints for the theatrical market. Thus, most of the costs of film production are incurred before there is any return on the invested money.

As Aphra Kerr notes, the same is true for video games. A game can spend several years in development, with a great deal of time devoted to its design, testing, and redesign.[37] Again, there is significant investment in the production of the game master, but the reproduction and distribution of the game are relatively cheap. Much like the film business, most of the money is spent before there is any return on the investment in the game.

In addition, both films and video games are high-risk businesses, and many films and games lose money once they are released. The breakeven point for films is highly contested, because the accounting practices of Hollywood studios are notoriously complex.[38] Studios will charge against production budgets the use of facilities that the studio owns, as well as administrative and overhead costs. Moreover, there are numerous sources of revenue for any given film, and films that perform poorly domestically can make up the difference in foreign exhibition markets.

Still, as Peter Dekom argues, profit margins are shrinking across the board, because the costs of film production are rising and the demands of profit participants (for example, actors and directors who take a cut of the gross box office receipts) are growing.[39] Therefore, the film business is increasingly risky, and it is not difficult to imagine how a film might lose money. Furthermore, when a tent-pole picture performs poorly at the box office, the disappointing returns have much broader impact because they erode a studio's ability to finance other productions.

For video games, the accounting may not be as complex, but the failure rate is still high. Kerr suggests that only 3 percent of video games actually turn a profit.[40] Admittedly, this figure seems rather low, but like film studios, video game publishers depend on successful titles to generate enough revenue to support the production of games that fail. When games are successful, they can see a return on their investment quickly: *Grand Theft Auto IV (GTA IV)* was estimated to have cost $100 million to produce, but it generated $500 million in sales in the first week of its release.[41]

GTA IV illustrates one of the most important links between the film and video game industries. Because both industries must contend with high up-front costs and a high risk of failure, both try to release titles that carry some promise of success. Therefore, both films and video games often follow popular generic conventions. For example, the action/adventure film genre frequently generates large gross receipts, and the top-earning films are often of this genre. Video games also have genres, such as first-person shooters (FPSs) and role-playing games (RPGs), and developers hold firmly to these established game genres because they have been proven in the market.[42]

Film studios and video game companies also attempt to secure success by developing franchises. When a film is popular, studios try to maximize the commercial value of that popularity by releasing sequels. Likewise, video game publishers release sequels to popular games that involve the same game characters (the *Tomb Raider* and

Xenosaga series are examples). Often, the newer versions of established games reproduce the gaming elements while employing completely new storylines and characters; examples include the *GTA* and *Final Fantasy* series. These franchises can also be an important point of convergence. When a film franchise is established, video games can be released with every new film, as in the case of the *Spider-Man* games mentioned earlier and the video games released for each *Lord of the Rings* film.

I should also note that, when it comes to marketing their products, film studios and game developers both target the same demographic. Males under the age of 35 are the largest target market for video games.[43] For films, the core audience is between the ages of 16 and 39, and most action/adventure tent-pole films primarily have a male audience.[44] Of course, there are exceptions, and females and older adults attend these films and play these games. Yet, sharing a similar target demographic means that these two industries share an understanding of the challenges they face in the marketplace.

Given that the film and video game industries confront similar challenges in the marketplace and that they meet these challenges in similar ways, it is not surprising that these two industries often work cooperatively. One way in which this cooperation is manifested is in the effort to work around different labor practices in the two industries. For example, while a good deal of the talent in the film industry is unionized, most of the video game production companies are not. The video game industry has built itself around a mythos of cultural and creative production, one that values individual ideas, encourages broad participation, eschews traditional corporate hierarchy, and rewards innovative contributions.

As Stephen Kline, Nick Dyer-Witheford, and Greig De Peuter point out, however, the reality is different from the myth.[45] For young people, and young men in particular, working for a video game company is often regarded as a dream job (anyone who has taught courses on video games to college undergraduates knows this to be true). Video game companies are well aware that they have access to a young and eager talent pool, one that is willing to work long hours for smaller salaries than those offered by other media industries.

This fact was made public by the infamous "EA Spouse" posting on *LiveJournal* in 2004. This anonymous post complained about the mandatory unpaid overtime that was demanded of game designers at EA Games. It was later revealed that the author of the post was the fiancée of an employee who was bringing a civil suit against the company. EA settled the suit out of court and reclassified 200 employees so that they were eligible for overtime but ineligible for stock options.[46] In other words, while the issue of labor conditions was addressed by EA, it was at the expense of the reward system that supposedly allowed video game workers to enjoy the financial success of the company. I should note that, while the suit was successful, it did not seem to result in any ongoing labor organization at EA Games, and in EA Spouse's original post, the word "union" does not even appear.

On the other hand, unions do have a presence in the film industry, and one of the most visible and active is the Writers Guild of America (WGA). In 2007–2008, the WGA organized a strike involving, among other things, issues related to writers' compensation for digital media, such as the rebroadcasting of television episodes online. Video games, as far as the WGA is concerned, fall under a separate category, and the Western branch of the guild has even formed the Videogame Writers' Caucus. The caucus includes among its goals the unionizing of video game writers, and it also has organized and promoted a video game writing award that is restricted to union members.

While reporting on the announcement of the award nominees, Eli Green, blogging for *Comic Book Bin,* expressed concern about unionizing video game writers and argued that a strike might disrupt the development of video games.[47] Of course, even during the writers' strike in 2007–2008, the WGA cut interim deals. For example, Marvel Studios cut such a deal, the details of which were not disclosed.[48] It is reasonable to assume that even if the WGA succeeded in its efforts to fully unionize the video game industry, similar agreements could be made, especially for studios like Marvel that are involved in both film and video game production.

The policies of the Screen Actors Guild (SAG) governing the use of SAG talent in video games provide a great deal of flexibility. The guild allows video game producers to opt for either a "Full Interactive Media agreement," which covers all video game productions, or a "One Production Only (OPO) contract that allows the producer the option of using SAG members for a specific game title. Producers can sign up to three (3) OPOs a year."[49] In other words, a video game company can use SAG talent on a specific project, like a film spin-off, without a long-term commitment to the union.

In addition, the current SAG contract for interactive media does not compensate talent with residual payments. As Ben Fritz, writing for *Variety,* notes, "SAG almost went on strike trying to get residuals for top selling videogames in 2005. They even staged a protest at Electronic Entertainment Expo (E3) that year. . . . SAG ultimately failed. . . . All it managed to get was a 36% increase in the base day rate."[50] Jonathan Handel, a Los Angeles attorney who specializes in entertainment and technology law, closely tracks the union's contract negotiations and those of the American Federation of Television and Radio Artists (AFTRA). He notes:

> AFTRA recently approved its interactive [i.e., video game] voiceover contract, while SAG voted down its similar pact, representing a rare defeat for SAG's new administration. . . . That means that video game companies can easily move over to AFTRA—or go nonunion. The hard reality is that neither SAG nor AFTRA control the labor supply in this area, leaving them little leverage in negotiations.[51]

Where issues of labor are concerned, unions that provide talent for the film industry have been cooperative with video game producers, in some cases allowing these producers to work with both union and non-union labor.[52] However, these contractual conditions are for baseline compensation and do not apply to the more bankable talent. When Robert Downey Jr. provided voice work for the *Iron Man* video game, his compensation was almost certainly different from that stipulated by the SAG contract. The use of film stars in the production of video games further illustrates the degree of collaboration between these industries, and this access to talent is an important advantage for the video game producers.

For film producers, the advantages of convergence can be found in the licensing agreements behind these video game spin-offs. The practice of product licensing is not new to the film industry: even under the old studio system, the likenesses of film stars were commonly used to promote products. Disney, in particular, has a long history of licensing its animated characters for a variety of children's toys, books, games, and records. *Star Wars,* however, is the film most often credited with making product licensing a standard Hollywood practice. As Al Ovadia observes, "[W]hen Star Wars–licensed products grossed over $1 billion . . . worldwide at the retail level in the first few years, the other studios took notice and began planning productions that would lend themselves to licensing."[53] Indeed, studios today actively recruit licensing tie-ins, sometimes before a film even begins production.

The particulars of agreements can vary significantly depending on the items produced and the popularity of the content. If a product is developed around an established and recognized character, such as Spider-Man, the terms of that agreement will most likely be different than for a film character who does not have a built-in audience. Furthermore, some established characters are more popular than others: Spider-Man is more popular than Daredevil, for example.

In spite of the variations in specific licensing agreements, most follow a general form. As Karen Raugust explains, the royalties collected on licensing agreements represent a percentage of the wholesale unit price, often in the range of 5–14 percent; the range for video games is generally between 8 percent and 10 percent.[54] There is usually a guarantee that a licensee agrees to pay the licensor either annually or over the license contract period, and this guarantee is based on the expected unit sales. Again, the range of guarantees varies significantly and is tied to the popularity of the content being licensed. Regardless of how a licensed product actually sells, the licensor still collects the royalty guarantee, and often a portion of this guarantee must be paid in advance when licensing contracts are signed.

As a result, when studios license their films for video games, they have little at stake, because studios make money from video game spin-offs even if the spin-offs fail to make money for their producers. Instead, the risk is borne by the licensee, in many cases the video game developers and/or publishers, and these companies even bear some of the costs when games are discounted at the retail level.[55] These publishers and developers also have a great deal to gain, however, particularly if they produce a video game based on a popular film, because the game will have a built-in market.

In addition, much of the work of the development stage is already completed for these games, which significantly reduces the cost for the video game company.[56] Video game companies are often given the film script, so a storyline does not need to be developed; the story is already established in the film, although games can deviate from the film's narrative, sometimes significantly. The game characters do not need to be developed or designed; indeed, frequently the film's actors will allow their likenesses to be used in the game design and will even do the voice-over work for the game.[57] Therefore, the game companies do not have to invest time and money to develop a brand-new game, which may or may not find a market. Instead, the licensing agreement gives these companies a set story, set characters, and access to film actors with established fan bases.

In addition to the traditional forms of content licensing, video games require another type of license. Video games operate on gaming systems, and game developers must design games that can operate on these systems. Currently, the most popular console systems are produced by Sony, Microsoft, and Nintendo, and each of these systems requires different programming.[58] In order to program for a specific gaming console—the Sony PS2 for example—the game publisher must pay a licensing fee to Sony, and when the game appears on store shelves it can then display the PS2 logo indicating that the game can be played on that particular system. Therefore, in addition to the licensing fees that must be paid to studios, companies that develop games based on film content must also pay licensing fees to the various console producers.

For a company like Sony, these licensing fees can be lucrative. Sony not only has a division devoted to the production of video game platforms (including the PS2, the newer PS3, and the handheld PSP), but it also owns film studios, specifically Columbia and MGM. When Sony released *The Da Vinci Code* in the summer of 2006, a video game was released for the PC, Microsoft's Xbox, and the PS2. Not only did Sony's film division collect licensing fees for all of those video games, its video game division also collected revenue for the PS2 license. While Sony is the exception rather than the rule, it provides an exemplar for the practices that can connect the film and video game industries.

In spite of the advantages of convergence, game spin-offs are sometimes of questionable quality and often fail in the marketplace. As Kerr notes, however, the majority of games fail in the marketplace— whether they are spin-offs or not.[59] Convergence, therefore, is a way for both film studios and game producers to hedge their bets and give their products an advantage. Furthermore, as opposed to other games, film spin-offs are not produced to function only as games; rather, they are designed to be part of an ancillary product campaign. Because the licensed product serves as a reminder of the film, in some ways a video game is no different from a promotional T-shirt or a fast-food souvenir cup.

There is, however, one important difference: unlike T-shirts and cups, video games are able to offer messages that are much more complex. In addition to the game play, many video games contain "cut scenes" that literally cut away from the action of game play. These cut scenes are often used to forward the storyline of a game (i.e., game characters discuss which enemies they have defeated and whom they must defeat next), and to that end they may resemble scenes from narrative films. Indeed, in some of the *LOTR* video games, actual scenes from the films are used as cut scenes to set up the action for the subsequent game play.

Sometimes, special features are unlocked by successful game play. For example, some of the video games based on Marvel Studios films contain Easter eggs of the original comic book covers depicting battles and villains that appear in the game; they may offer additional promotional texts as rewards for successful game play as well. Finally, these video games often require the prolonged engagement of the game player—some games can take up to 40 hours to complete. These games are complex texts, and this complexity allows for a persuasive function.

Studying Video Games

For many years, video games have been studied for their effects, specifically for the ways these games encourage or impart violent behaviors. This is particularly true in the field of communication, where many of the published studies have been concerned with how video games stimulate aggression and violence.[60] This particular approach to video games has gained a great deal of attention among politicians and social conservatives who view games as part of the broader cultural decline in the United States.

Over the last few years, however, media scholars have begun studying the potential benefits of video games as well. Many of these scholars have been influenced by cultural studies approaches that view popular culture as an opportunity for empowerment and pleasure. Henry Jenkins perhaps best represents these scholars, and his work focusing on the pleasures of consuming popular culture includes video games among its objects of study. In important ways, Jenkins's approach to video games provides a response to the media effects research that has represented these games as potentially dangerous. In his book *Fans, Bloggers, and Gamers,* Jenkins provides a critique of effects research on video games and presents testimony he gave before Congress defending video games, and challenging the assumptions of effects research.[61]

It is easy for me to empathize with Jenkins, because I currently live in a state where our recently impeached governor, Rod Blagojevich, once took up the battle against violent video games, championing legislation that was ultimately deemed to be unconstitutional. Blagojevich repeatedly argued that these video games instilled violent behavior in young people. In light of these political circumstances, it is reassuring to know that scholars like Jenkins are working to garner some respect for video games as a medium. Similarly, in the Serious Games movement, both scholars and industry representatives study the use of video games as training and educational tools.[62] While these studies also examine the effects of video games, they do not proceed from the assumption that these effects are always negative; instead, these scholars are finding benefits in the practice of game play. Although video games have been viewed with suspicion over the years, this research program views video games more positively.

While I applaud research that views video games positively, I suggest that video game scholars should not eschew more critical perspectives, particularly when a critical perspective is warranted. For example, video games that are spin-offs of studio films are commercial products, and they serve an economic function. Video games *are* pleasurable; so are films. Indeed, the appeal of the video game spin-off resides in the assumption that playing the game will extend the pleasure of the film. Yet these video games also involve financial transactions in which money flows from consumers to film and video game producers.

The critical perspective I will use in this book is informed by the political economy approach to the study of media, a program of study with a long history dating back to Theodor Adorno and Max Horkheimer's work on the cultural industries.[63] The political economy approach is often criticized for imagining media consumers to be cultural dupes, passively absorbing the messages and ideology perpetrated by the media. In contrast to political economy studies, the cultural studies approach theorizes media consumers to be actively resistant to ideology and willing to negotiate different interpretations of media messages.[64] Some scholars of video games have argued that because video games are interactive, the video game player is similar to the active media consumer imagined in cultural studies approaches. Therefore, the political economy approach has often been placed in opposition to the study of consumer uses and pleasures, and Jenkins makes this opposition explicit in the conclusion of *Convergence Culture* when he refers to political economy scholars as "critical pessimists."[65]

There is, however, a new interest in issues of political economy where the video game industry is concerned. Kline, Dyer-Witheford, and De Peuter have authored an extensive material history of the video game industry.[66] In 2006, Kerr, in *The Business and Culture of Digital Games,* also argued for the importance of studying the video game industry as a cultural industry.[67] In the inaugural issue of *Games and Culture,* a journal devoted to the study of video games, Toby Miller authored the lead article and spoke of the need to address the issues of cultural production as they relate to the video game industries.[68]

While these material analyses look at the video game industry generally, this book will focus on a specific aspect: the convergence of the film and video game industries.[69] This convergence can best be understood as a form of media synergy. As Joseph Turow observes, synergy is a practice in the media industry that emerged with the deregulation of media ownership and the aggressive acquisition of smaller media concerns by larger media companies. Turow notes that many media outlets are now consolidated into a few conglomerates that are vertically integrated and highly diversified.[70] Sony, of course, is one of these conglomerates, as are Fox, Viacom, Disney, and Time Warner. Each of these companies has interests in film production, and each also owns a variety of media outlets in television, music, newspaper, and magazine production. Specifically, synergy is achieved when these companies repurpose content across a variety of different media, or windows of distribution; the film franchise is a prime example. Admittedly, synergy is an overused term, but film franchises are still key to the business strategy of the major studios, and video game spin-offs are important tactics in the franchise strategy, for good reason.

Like the film industry, the video game industry has become increasingly consolidated. On the hardware end of the industry, the consolidation is a reaction to market forces. The market will support only a limited number of game consoles, and in the twenty-first century the dominant three companies (Microsoft, Sony, and Nintendo) provide players with about as many options as the market can sustain. Software, on the other hand, is a market that can support numerous companies and many small game developers. Yet even the software market has experienced consolidation, best represented by the acquisitions of EA Games.

EA Games describes itself on the corporate web page as "the world's leading interactive entertainment software company."[71] It could just as well describe itself as one of the largest, and EA Games seems determined to get larger. In early 2008, EA Games acquired the VG Holding Corporation, which allowed EA to add BioWare and Pandemic Studios to their label. While both of these companies still publish under their own brands, their acquisition means that EA Games now owns several *Star Wars* titles, the *Baldur's Gate* series, and *The Lord of the Rings: Conquest.* These acquisitions are in addition to several others made by EA Games, and it has attempted to take over both Ubisoft and Take Two. Although these last two takeover attempts appear to have failed, they illustrate the aggressive desire of EA Games to acquire and grow.

EA Games is not the only video game company that harbors this desire. In 2008, Vivendi Games, which owned Blizzard and the highly successful *World of Warcraft* franchise, merged with Activision to form Activision/Blizzard. According to Adam Satariano, writing for www.bloomberg.com, since that merger the new company has now surpassed EA Games as the world's largest game publisher.[72] Satariano goes on to note that Activision/Blizzard has a good deal of cash on hand and is well positioned to acquire struggling companies in the current economic climate.

Don Reisinger, a blogger for *CNET,* offers this assessment of the consolidation on the software side of the industry:

> If we consider Hollywood—the model to which the video game industry is always compared—it doesn't take long before we realize that it's dominated by a handful of studios that effectively control a large percentage of the industry, while the independent studios are left trying to defy the percentages and get their innovative and artistic films to the masses. Since most fail, it's the big studios that enjoy profits as the independents try to find some way to stay alive. Who can say that this isn't where the video game industry is headed if companies like Activision and EA continue their acquisition frenzy?[73]

Who can say indeed? What can be said with certainty is this: when film studios work with video game producers, it is often one major, multinational conglomerate working with another. Because of their size and their global presence, each company has some understanding of the way the other does business. One thing that both film studios and video game companies understand is synergy, because it is a business strategy born of exactly this type of industrial consolidation.[74]

Synergy encompasses a variety of cross-promotional practices, and video game spin-offs often reflect these practices. For example, *The Lord of the Rings/The Return of the King* (*LR/RK*) video game contains several Easter eggs, including a segment called "Hobbits on Gaming" in which actors from the film discuss the merits of the video game that is being played. This segment links the experience of the film to the experience of the game in literal terms, with the actors attesting to the fact that the game play replicates the action found in the film. In other words, this game and others like it can operate as persuasive texts that promote the interests of media producers. Therefore, I will be treating video games as rhetorical texts. This approach is rather new in relation to both rhetorical studies and video game studies. The study of rhetoric, particularly in the United States, has traditionally focused on public address, with rhetorical studies of traditional media having also been widely accepted in the field. In the twenty-first century, however, some scholars have become interested in the rhetorical function of video games and have begun to view them as rhetorical texts.

Rhetoric and Video Games

Ian Bogost has argued that the real persuasive power of video games lies in the way games are designed, or what he calls *procedural rhetoric:* "Procedural rhetoric is a general name for the practice of authoring arguments through processes. . . . Arguments are made not through the construction of words or images, but through the authorship of rules of behavior, the construction of dynamic models."[75] Game design establishes rules that dictate how a player can interact with the game environment and other characters in the game and how game play is rewarded. The concept of procedural rhetoric is useful to the study of spin-off video games in that a game's design can reflect important elements of the film. Bogost offers as an example a video game that depicts the sport of "quidditch", which appears in the *Harry Potter* films and books.

Another example can be found in the *Fantastic Four* video games. In these games, the player has the option to play as any of the four characters from the film. Yet, in order to complete certain tasks and advance in the game, the player must draw on the special powers of each of the four characters. This design requires the player to use all four characters and thereby reinforces the theme of unity that underlies the *Fantastic Four* film (2005). Similarly, in *The Godfather* video game, players advance in the game much more quickly if they participate in a variety of side missions, such as extorting money from various businesses by promising the protection of the Corleone family. Some of these missions require extreme violence. By rewarding the player for completing these tasks, the procedural rhetoric of the game encourages the player to adopt a "mobster" mindset and become involved in some of the violent actions depicted in the film.

While these game designs are an important element, other visual and narrative aspects of game play are important as well. As Bogost points out, *Quidditch World Cup* is a textured game with details that appear in neither the books nor the films and with rules that promote a sense of individualism that differs from the values offered in the films. Still, this video game depicts Harry Potter in a way that resembles the young actor Daniel Radcliffe. In addition, although *Quidditch World Cup* is not a spin-off of the actual *Harry Potter* films, other games are, and of course, these games reflect the titles of the books as well. It is also interesting to note how these games visually depict a maturing Radcliffe in a manner similar to the later films in the franchise.

Although he does not identify it as such, Bogost's procedural rhetoric reflects a *ludologist*'s approach to video game studies. Ludologists are game scholars who draw on a tradition of game studies that focuses on the act of play; they view play as an activity that can be distinguished from other activities. More specifically, they are interested in how the rules and goals of games create special places or, to use Johan Huizinga's phrase, "magic circles."[76] Ludologists study video games *as* games. For example, Espen Aarseth is noted for arguing that game studies should focus on game play rather than on the textual elements associated with narrative studies.[77] Admittedly, Bogost's approach can be more practically applied than those of most ludologists, and his focus on the rhetorical functions of games links them to concerns and contexts that reside outside the magic circle of game play. Yet his theory of procedural rhetoric focuses on game play and game rules, and he maintains that the procedural elements are what make video games uniquely rhetorical.

As opposed to ludologists, *narratologists* view video games as narrative texts, and their analyses draw on a tradition of literary theory and criticism. Of course, video games are visual, and some scholars have argued that these games are perhaps best studied from the perspective of narrative film. For example, Mark J. P. Wolf provides a detailed analysis of the cinematic elements found in video games and identifies how the narratives of video games often reflect the elements of narrative film.[78] These parallels are difficult to ignore because video game spin-offs parallel, intersect, and extend film narratives in a variety of ways. Some even reflect actual scenes from the original film, as in the *LR/RK* video game, in which game play is digitally mapped onto actual scenes from the film. Other games tell an alternative narrative that parallels and intersects with the narrative of the film, such as the story of Sora, the main character in Disney's *Kingdom Hearts* video games. In these games, Sora's efforts to return home require him to help many characters from different Disney productions (Hercules and Tarzan, for example), so his story intersects with their narratives and the original films. Finally, some video games reference narratives that serve as source material for theatrical films: in the *Spider-Man* games, characters from the films appear, as do characters that are found only in the comic books, positioning the games as points of connection between the films' narratives and the original source material. For example, in the *Spider-Man 2* game, "Spidey" flirts with Black Cat (a character from the comic books who does not appear in the films) and works with her to solve some of the game's missions.

When games are produced as part of film franchises, the design of the games can interweave game play with film narratives and supplemental messages, and game play often advances a narrative that links the game to the film. In addition, these games can be designed so that game play opens up a variety of promotional texts that extend the experience of the film. Therefore, for my critical purposes, it would be unwise to separate the game design from the game narrative, the visual representation of the film's characters, or the other texts and messages contained in the game. The narratives and the texts in these games are already inscribed into the software of the game, but it is through the interaction with the game player that these narratives are advanced and these texts are activated and experienced.

To understand how a player's interaction with a game can be directed in this manner, it is necessary to examine interactivity from a different perspective. It has often been argued that the interactive nature of game design actually empowers players with a sense of agency not found in the traditional forms of media, specifically film. For example, Wolf observes, "Rather than merely watching the actions of the main character, as we would in a film, with every outcome of events predetermined when we enter the theater, we are given a surrogate character (the player-character) through which we can participate in and alter the events in the game's diegetic world."[79] Jo Bryce and Jason Rutter make a similar observation, contending that this interactivity places the player in a powerful role that is distinct from more passive forms of film reception.

Bogost challenges the notion that interactivity imparts player empowerment, arguing that "sophisticated interactivity can produce an effective procedural rhetoric."[80] He goes on to argue that interactivity carries an *enthymematic* function. The enthymeme is an Aristotelian concept that refers to a deductive argument that requires audiences to complete the reasoning process. Often, an enthymeme omits a premise, and it is up to the audience members to fill in the missing information by calling on knowledge that they share with the rhetor. In this manner, an enthymeme is interactive because it requires the audience to participate in the persuasion process. Just as the enthymeme requires the participation of the audience, interactivity invites video game players to participate in the persuasive practices built into the games. Indeed, a video game player's interactivity becomes enthymematic to the degree that play becomes a means of accessing and activating persuasive messages and procedures that augment promotional practices. In other words, through game play, the game player actively participates in the persuasion process.

Playing the Hollywood Games

In this book, I will look at film spin-off video games as rhetorical texts designed to serve the interests of production. Rhetorical criticism as a method is not as concerned with the actual effects of discourse on an audience as it is with the way messages address a rhetorical exigency or context. I will thus make no claims about the effects of the rhetorical strategies I observe, but will instead anchor my observations in specific contexts of production as I explore various cases in which film content has been repurposed as video game content.

I observed all of the games I analyze being played to their conclusions and won, and I took notes on the games in the process.[81] These notes formed the foundation of my analysis, while the strategy guides that have been published for these games provided supplemental texts. Like the games, these guides are commercially published and are part of the licensed product mix of the film franchise, and they often include additional messages that serve a promotional function. In order to keep my practices consistent across these games, all were played on the PS2 platform, which has been a dominant format for some years and enjoys a large consumer base. Finally, I chose games that reflect different contexts; these choices are representative of the practices of film and video game convergence, but they are not exhaustive.

This book is not the first attempt to study the relationships between video games and films. The anthology *Screenplay* was devoted to essays that examined video games from the perspectives of film aesthetics, spectatorship, and narrative.[82] The persuasive function of these games, however, has not been examined, nor have they been studied in a manner that links them to the context of production. In the chapters that follow, I will show how these video games do more than promote specific films; they incorporate game play in narratives and messages that address problems associated with media brands and media franchising.

In the next chapter, I will extend the critical engagement of the concept of interactivity introduced in this chapter. Specifically, I will demonstrate the rhetorical function of interactivity through a close analysis of the video games released in tandem with the films of *The Lord of the Rings* trilogy. As I indicated earlier, these games are designed to expose the player to a variety of promotional messages. I anchor these messages in the efforts of the films' producers to reach fans of the J. R. R. Tolkien books by offering the trilogy of films as a way of experiencing Middle Earth. These messages attempt to position the games as extensions of the film experience, equal to if not better than the films themselves. My analysis in this chapter will illustrate how a video game can be designed rhetorically to extend the experience of a film and augment the practices of the film franchise.

In the third chapter, I will discuss the emerging practice of taking an older film's content and releasing it as a video game. Examples of this practice include video games based on the films *Scarface* (1983), *The Warriors* (1979), *Ghostbusters* (1984), and *The Godfather* (I will focus specifically on the latter's video game). Francis Ford Coppola refused to participate in the development of *The Godfather* video game, and his lack of involvement invites a reconsideration of authorship. Although Coppola's authorship is closely associated with *The Godfather* franchise, the game allows the player to construct a new narrative that intersects the films' narratives but stands apart from Coppola's own. In this way, the game player's agency supplants Coppola's and thereby fulfills the interests of a studio that has had to confront an uncooperative director.

In the fourth chapter, I look at games released by Marvel Studios, a company designed to turn Marvel's comic book characters into feature film heroes. Although Marvel Studios began producing films only in the mid-1990s, its success has been historic; its *Spider-Man* franchise alone is responsible for 3 of the top 20 domestic grossing films of all time.[83] Marvel Studios has also had success with the *X-Men* films, the *Fantastic Four* films, and *Iron Man* (2008). In this chapter, I will look at several games that were released as spin-offs of these films. My analysis will demonstrate that these games used strategic intertextuality to link the games to the experience of the films and to the comic books on which the films are based. These games remind players that Marvel's films are true to their source material: the original comic books. In many cases, the games serve as intertextual points through which the narratives of the films and the narratives of the comic books intersect.

In the fifth chapter, I will analyze the two *Kingdom Hearts* video games, which Disney produced with Square Enix, the video game company responsible for the popular *Final Fantasy* franchise. While the *Final Fantasy* games have enjoyed popularity in the United States, they are even more popular in Japan, where they originated.[84] The *Kingdom Hearts* games include several Disney characters and draw from many of Disney's animated films, but they also include characters from the various *Final Fantasy* games. In these games, the player leads the character Sora on missions in which he helps Donald, Goofy, and "King Mickey" fight the Heartless, who are minions of darkness that threaten the Magic Kingdom. I will show how these games function as cultural hybrids, weaving into a single narrative characters that are strongly identified with both American and Japanese popular culture. I will argue that this particular game narrative functions as a metaphor for the negative aspects of media globalization, a narrative in which, ironically, Disney becomes a champion against these negative aspects.

The final chapter will look to the future and identify key issues that should be pursued in the further study of film and video game convergence. I begin in the present, however, by analyzing how the advancement of game technology is playing out in the context of convergence. Specifically, I will look at the different business strategies behind Sony's PS3 and Nintendo's Wii and the different fortunes of those consoles. I will look at the phenomenon of online gaming and massively multiplayer online role-playing games (MMORPGs) and discuss the conflicts that can arise between players and game companies. The film industry has a long history of collaboration with the military, and that collaboration has carried over to the video game industry. Therefore, I will look at the connections between the film industry and the military and discuss how those connections are shared with the video game industry, and how they might be examined in future convergence studies. I conclude that, as long as convergence offers advantages to the participants, it will be an ongoing practice that will determine the types of media products available to consumers.

TWO

PLAYING THE GAMES, BEING THE HEROES

When he stood onstage at the 76th Academy Awards, Peter Jackson was clearly a happy man. Not only had he won honors for best director, but his film *The Return of the King* (*RK*, 2003), the last in *The Lord of the Rings* (*LOTR*) trilogy, also won best picture. New Line Cinema, the studio that produced the trilogy, had to be happy as well: the combined box office revenues of the three films exceeded $1 billion.[1]

Both Jackson and New Line had come a long way to arrive at this point. Jackson began his career in the horror/comedy genre with such films as *Bad Taste* (1987) and *Braindead* (1992). He received his first Oscar nomination for the film *Heavenly Creatures* (1994) and was nominated for both the first and second films in the *LOTR* trilogy, *The Fellowship of the Ring* (*FR*, 2001) and *The Two Towers* (*TT*, 2002), before winning for *RK*. New Line began as a distribution network for the college art film circuit. The company handled mainly foreign and art films but also achieved success by rereleasing the camp classic *Reefer Madness* (originally released as *Tell Your Children*, 1936) and by first distributing and then producing the films of cult favorite John Waters. In 1984, New Line released *A Nightmare on Elm Street* (1984), which went on to become one of the studio's first franchises.

Although New Line had experience with producing and marketing film franchises, *LOTR* was a major step forward for the studio. In fact, the project would have been incredibly ambitious even for a major studio like Paramount or Disney, and it was even more so for a "mini-major" like New Line. When Jackson first approached

New Line with the idea of bringing the trilogy to the big screen, he pitched a one-picture deal. New Line countered by offering Jackson a contract to write, direct, and produce films for each book in the trilogy. In an unprecedented move, New Line put all three pictures into production simultaneously. Thus, from its inception, *LOTR* was conceived as a franchise and produced as a multipicture project.

In addition to the nominations and awards, the *LOTR* film trilogy became an economic force of some magnitude, and the films' presence in the global economy makes them part of an immense media franchise. Although the films could count on the popularity of the Tolkien books, other media and ancillary products helped to spread the trilogy globally. These products included action figures, books about the production of the film, soundtracks, and even repackaged books of the original trilogy with covers based on stills of the film. Like most media franchises, video games were part of the ancillary package developed for the *LOTR* films. While most franchises release new games with each new film, games were only released for the second and third films in the *LOTR* trilogy. Still, these games generated significant revenue, and the EA Games 2004 Annual Report listed both games among the company's "platinum titles."[2]

The *LOTR* video games provide an example of the interesting vagaries of licensing practices. Vivendi Universal Interactive was given the original license for *LOTR* video games and went on to release video games of *The Hobbit* and *FR*. Vivendi Universal, however, confronted some significant problems, and it released no other *LOTR* games. First, parent company Vivendi began experiencing financial trouble and sold off 80 percent of Universal to GE in 2003, creating what is now known as NBC/Universal.[3] These financial problems and the subsequent split may have disrupted any plans to continue with the *LOTR* license. Second, New Line acquired the film rights to the books and was able to license its own video games based on the films—which it did, contracting with EA to produce the games for *TT* and *RK*.[4] EA Games would also release two more games for the *LOTR* franchise: *The Third Age*, a role-playing game, and *The Battle for Middle-Earth*, a real-time strategy game. Although these games include a number of elements from the films (*The Third Age* game has several clips from the trilogy), neither is actually a spin-off of any of them. As discussed above, the *FR* game produced by Vivendi

Universal was not even part of the New Line franchise. I will therefore focus my analysis on the *TT* and *RK* games.

Both of these games were released before their respective films: *TT* was released on October 21, 2002, two months before *The Two Towers* was released on December 18, 2002. *RK* was released on November 2, 2003, more than a month before *The Return of the King* hit theaters on December 14, 2003. Although releasing the games before the films may seem backward, the strategy assured that the games were readily available for the Christmas buying season, and many of those who saw the films on their release dates found the games under the Christmas tree in a matter of days. This temporal association between the games and the films was augmented by campaigns that connected game play with the experience of the films, and the ads for *RK* invited potential players to "Play the Movie. Be the Hero."[5] Many consumers accepted this offer, and the game sold over 3 million units. *TT* did even better with sales exceeding 4 million units.[6]

As discussed in the first chapter, the connections between film and video games have not escaped the attention of scholars in the emerging study of video games. Although a good deal of scholarship on video games conceptualizes game play as a liberatory, interactive experience, I will take a more skeptical view. I will argue that the construct of "interactivity" must be reconsidered in order to comprehend how video games connect players to the economic interests of production. I will begin by reviewing the literature on interactivity in video games and by questioning the tendency of video game scholars to associate interactivity with agency and ideological resistance. I then will provide analyses of the *TT* and *RK* games, demonstrating how their structures use interactivity as a means of exposing game players to messages that link them to the *LOTR* experience.

Interactivity and the Game Player

In the study of new media generally and video games specifically, the concept of interactivity figures prominently. Although definitions of interactivity vary, where video games are concerned the concept is commonly used to refer to the game players' agency. As Matt Garite notes, "This emphasis on the active role of game players is a common trope that appears repeatedly in discourses on interactivity. . . . Game players are thus seemingly granted a degree of agency and choice."[7] Ben Sawyer, Alex Dunne, and Tor Berg claim that "the notion of interactivity means that the decisions and skills of the player will move the story in a certain direction," thereby allowing the player to actively change the game as it is played.[8]

Many of these changes are brought about by the player's use of an avatar. As Miroslaw Filiciak observes, some games allow players to create their own avatars and form their own identities through interactive game play.[9] The concept of agency itself is deeply rooted in the mystique of the video game avatar. As Bob Rehak observes, "[P]art of what users seek from computers is continual response to their own actions—a reflection of personal agency made available onscreen as surplus pleasure."[10] In other words, the pleasure of playing video games lies in the ability to create or change your avatar and thereby change your experience and the progression of the game.

The association of agency with interactivity is heightened when scholars theorize the variety of relationships that spectators have with films and that players have with games. Wee Liang Tong and Marcus Cheng Chye Tan suggest, "Playing the game resembles watching a film. . . . [The] difference lies in the levels of interactivity offered by the game, between the gamer and the game-environment."[11] Interactivity, then, is not just a means of character control; it is also a way for game players to construct their own environments and narrative spectacles. Accordingly, Jo Bryce and Jason Rutter contend that this interactivity places the player in a powerful role, which is distinct from more passive forms of film reception:

> Film audiences have a history of being viewed as gatherings of passive individuals who sit, in a darkened cinema, as the light and sound of the cinema projection pours over them. In this environment audience members are "passive" recipients of the narrative of the film. . . . As the previous discussion has shown, game players or audiences are more actively engaged than film viewers in both the narrative and the other events within the game environment. The ability to modify both of these aspects of a computer-based game shows a level of interaction with the text that is not provided by traditional cinema or Hollywood blockbuster movies.[12]

In conceptually linking interactivity with player agency and juxtaposing this agency with the passive cinema viewer, Bryce and Rutter suggest that the video game player is similar to the active viewer theorized in some cultural studies and in television scholarship.[13] These studies have found that television viewers actively bring meaning to the programs they view, often negotiating and even resisting the ideological messages conveyed in television programming. This association is understandable to the degree that video games, particularly those played on game consoles, are often enjoyed through televisions. Yet, equating the video game player with the actively resistant television viewer is problematic for several reasons.

First, the agency attributed to the active audience discussed in television studies stems from ideologically resistant, interpretive practices that occur when viewers actively bring their own meanings to texts. The agency attributed to the interactive game player, however, has little to do with interpretation and everything to do with game play. Although a player may be able to change a character in the game or alter the direction of the story, these actions do not necessarily denote ideological resistance. As Mark J. P. Wolf has noted, while video games are interactive, the interaction is directed toward some goal or motive that makes the game worth playing and winning.[14] Interactive video games are still games, and they have a prescribed set of rules that regulate how the game is played. The player who resists the rules of the game, or ignores or reinterprets them, is most likely to lose the game. In other words, video games reward compliance. More to the point, successful interaction in the context of game play (i.e., winning the game) is not a practice that brings meaning to the text, but rather one in which the player follows the text closely.

Second, the video game player is subject to the limitations of a game's structure—limitations that are predetermined and imposed by the game's designers. Rarely do the choices in video games significantly change the structure of the game; rather, they might shift the specific details of a character's attributes or the scenic features. While some video games do allow the player a great deal of latitude in the designed game play, in the case of commercially developed and manufactured games—the ones that drive the video game industry—the player is rarely given the agency to change the game's structure or design. Game manufacturers have a vested interest in creating and marketing a specific game experience; by disrupting this experience, such player agency would actively undermine the purpose of mass producing and marketing a uniform product. Limitations must therefore be imposed on the choices a player can make, and these limitations do not always allow for the kinds of changes that could be equated with ideological resistance. A player can resist the content of a game, perhaps questioning the way women are portrayed in a game, for example, but that type of reading practice extends across a variety of media and is not necessarily a product of video game interactivity. The controversy over *GTA: San Andreas* was a case in point. If players activated a particular "cheat," they could unlock a hidden scene in which a woman performs oral sex on their avatar. The scene is produced by interacting with the game, but this interactivity hardly signifies a political response challenging the stereotypical representation of women as sexually subservient to men.[15]

Third, in many games some of the most important elements are *not* interactive. The levels of many video games are punctuated with cut scenes, cinematic video clips over which the player has no control. After a level is played, the player views a scene in which there is no interactivity, and the player assumes the passive posture associated with cinema spectators, as described by Bryce and Rutter.[16] Many cut scenes are familiarly cinematic in appearance and provide the player with information that helps to advance the game and move the narrative forward. Consequently, at the points in the game where the narrative is advanced, the player is also rendered passive and has no agency in changing the game's narrative. As Chris Crawford has noted, truly interactive narratives are an ideal seldom, if ever, realized in video games.[17] Given that the reward for completing a level is a cut scene with a predetermined message, the interactive aspects of many videos games do not empower players to resist the messages that are woven into a game's structure. More to the point, those elements of a game that can convey ideological messages are also the elements in which the player enjoys the least agency.

Rather than associate interactivity with agency and resistance, I propose a different approach that can critically engage the economics of the *LOTR* franchise. For example, P. David Marshall observes that, while interactivity can refer to active consumer practices, it can also signify the industry's strategies that are used to attract audiences in a media-rich environment.[18] He specifically identifies video games as part of these strategies, arguing that video games serve as a metaphor for the way the cultural industries connect with audiences and keep them "within the system of entertainment choices."[19] Following Marshall's reasoning, the interactivity offered by video games functions as a means of incorporation. As Ien Ang has observed, the plethora of media now available to the consumer allows for many choices and alternatives for entertainment, and this freedom of choice is often portrayed as agency. However, she notes, the representation of "choice" as agency legitimates the expansion of new media technologies and obfuscates the interests of the industries behind these technologies: "Seen this way, the figure of the 'active audience' has nothing to do with 'resistance,' but everything to do with incorporation."[20]

Therefore, I propose that interactivity be theorized as a form of incorporation and that the structure of a video game be viewed as something that creates an interactive experience that connects players to the interests of production and attracts them to a particular media choice. More specifically, I will argue that the structure of *TT* and *RK* connects players to the *LOTR* franchise and the mythic experience of Middle Earth. This approach, while acknowledging that the game player is not passive, recognizes what Lawrence Grossberg has noted: "People live their subordination actively; that means, in one sense, that they are often complicit in their own subordination, that they accede to it."[21] I maintain that video game play, with its system of rewards and messages, creates a structure in which players accede to their investment in the games, the films, and the fan culture of *LOTR*.

Of the millions who purchased these games, undoubtedly many were part of the *LOTR* fan culture. In his book *Textual Poachers,* Henry Jenkins describes how fan culture has become a mainstreamed demographic for cult texts and how fan culture creates a mythos around texts that reside slightly outside the conventional media output. These cultish texts are treated by fans "as if they merited the same degree of attention and appreciation as canonical texts" and are given the type of close reading usually reserved for "high culture" texts.[22] Fans are important marketers of these cult narratives, not only because fans enjoy and purchase materials surrounding the main text (books, ancillary products), but also because fans disseminate these texts to other fans. Taking his analysis of fan culture one step further, Jenkins argues that fans move beyond simply enjoying the texts to become co-creators of those texts. Fans, he claims, "raid mass culture, claiming its materials for their own use, reworking them as the basis for their own cultural creations and social interactions."[23] Using the classic example of *Star Trek,* Jenkins examines both how fan culture digests the stories and disseminates the materials and how the *Star Trek* franchise encourages fans themselves to create new texts, such as fan-authored fiction. Fans interact with the materials of cult texts to "integrate media representations into their own social experience."[24]

The *LOTR* narrative itself is a cultish text that has garnered its own fan culture. According to one of the documentaries on the DVD containing the extended edition of *FR,* "it is the second most-read book of the 20th century, after the Bible," and its fan base is unusually persistent in memorizing the details of and searching out errata in the Tolkien world.[25] This careful attention to the text of *LOTR* makes members of this fan culture a built-in audience for the films and prime candidates to purchase video game adaptations of the films. Indeed, New Line was probably banking on the devotion of *LOTR* fans when it decided to produce all three films simultaneously. In addition, the fan culture surrounding *LOTR* provides an important example of why interactivity needs to be reconsidered when explaining video game play. These fans have actively produced new texts, developed new intertextual relationships, and appropriated the cultish texts of the trilogy, but all of these active practices are in sharp contrast to the interactivity of the game play in *TT* and *RK.*

I will demonstrate that the interactivity of these games is structured in such a way as to augment the synergistic relationships of the *LOTR* franchise. More specifically, I will argue that the player's interaction with the games is rewarded with messages that link the games with the films. Therefore, the interactivity of the games draws players in while constantly relating the experience of game play to the experience of the theatrical films, allowing the fan of *LOTR* to become more closely connected with the film franchise. In my critique of the games, I will demonstrate how the games' tutorials introduce players to the juxtaposition of interactive game play and promotional reward. I then show how players are rewarded for beating levels with more of these promotional messages. Such messages, which are not created by the player but are activated by playing the game, thereby turn interaction into incorporation.

Playing the Game, Playing the Movie

Although video games are interactive, it is interesting to note that these particular games begin by rewarding passivity. Once the game discs are inserted, both games begin with the on-screen logos of EA Games and New Line Cinema. New Line's opening logo is identical to that which precedes its theatrical releases, recreating a visual experience similar to that of sitting in a theater. These logos are followed by trailers for the games depicting scenes from both the films and the games. These trailers play automatically if the player does not respond to the game interface and begin game play.[26] The scenes in these trailers depict Frodo Baggins escaping yet another harrowing situation, Samwise Gamgee running through mountains and caverns, and Gandalf the White obliterating evil orcs and goblins by the hundreds. All these scenes are, or at least will become, familiar to the player after the games are completed, as some of the footage is taken directly from the cut scenes of the games, while others are based on actual game play scenarios.

The practice of using a trailer to introduce a game is not new, nor is it unique to these games; *Final Fantasy X*, *Xenosaga*, and *Star Ocean* all use trailers in their introductions. Unlike these three examples, however, the *TT* and *RK* game trailers enjoy a direct relationship with major theatrical films, and the trailers certainly exploit that relationship. For example, the end of the *RK* trailer displays the title *The Lord of the Rings: The Return of the King*, while the character Gollum (Andy Serkis) from the film rants, "We swears to kill master. We swears on the precious." Thus, the *RK* trailer is not only a conglomeration of elements from the film and the game, but also a microcosm of the complex intertextual relationship between the two. The trailers introduce both the game and the type of synergistic messages that will punctuate the game play.

Both of these games begin with narration provided by Cate Blanchett as the character Galadriel, in which she explains to the player "the story so far." As these games are tied only to the second and third parts of the trilogy, retelling the backstory becomes necessary in order to make the plot understandable. Ironically, Galadriel as a character is hardly present in these games, yet her role in the narrative of the films is one of utmost importance: her character is immortal and exists outside of time, and thus she serves as the omnipresent narrator of the films. Her appearance in the games reflects this omnipresence, connecting the video games to the film narrative, and her narration is used to verbally fuse visual elements culled from the films and the video games. Within the games, however, Galadriel is not really omnipresent; her presence is limited to her synergistic function.

Because EA did not release a game for the first film, a great deal of footage from *FR* is incorporated into the introduction to the *TT* game. This footage is rather seamlessly interwoven into the computer-generated imagery (CGI) of the game and calls attention to just how much CGI was used in the film. The first scene in the game is the battle of the Last Alliance as depicted in the film, and it transitions into actual game play as a tutorial session in which the player learns the controls for the game. Given the complexity of video games, part of the challenge of a game is learning how to control the avatars. To this end, as players begin the actual game play, the system uses built-in controls to instruct the player as to which buttons to press.

The tutorial restricts players' interactivity, limiting them to one character and even prompting them to make specific moves. After this first session, the game CGI transitions back into the film footage, showing Isildur slicing the infamous ring from Sauron's hand. In this way, the player becomes part of the battle that begins the *LOTR* trilogy, and the player's introduction to the game mirrors the first film's narrative introduction. While this session is interactive, the player enjoys little agency; after a set amount of time, the tutorial ends and a cut scene from the film begins automatically, regardless of how the player performs in the tutorial.

The *RK* game also begins with an important battle. This time, the tutorial is conveyed in a cut scene culled from the end of the second film. In this scene, the "Battle of Helm's Deep," the heroes engage in an intense, climactic struggle to maintain their stronghold in a final defense of the kingdom of man against their enemy: Saruman and his minions. The defeat of Saruman at Helm's Deep signals the turning point of the series; all the elements are now in place to begin the offensive against Sauron, which will be the narrative of the third and final film.

In the video game, however, the "Battle of Helm's Deep" has a specific, concrete purpose—as a tutorial. In this "training" mode, the player is given the avatar of Gandalf, while another character from the film, Gimli (John Rhys-Davies), yells instructions to the player. Interestingly, Gimli's dialogue is most distinct from the rest of the dialogue in the game, as it is only related to the game play and not to the film itself. The rest of the game's dialogue, recorded by the actors specifically for the game, consists either of lines taken directly from the film ("Run, Mr. Frodo!") or close approximations of the film's dialogue. When Gimli yells instructions at Gandalf, however, he uses terms from the video game world: "Use fierce attack!" "Don't forget to block!" "Speed attack works best!" Although Gimli is supposed to be shouting at Gandalf, he is actually shouting instructions to the player.

After Gimli trains the player in various attack moves, a new cut scene appears, taking the action up to the walls of Helm's Deep. The game introduces new types of enemies, new types of fighting, and more training. Again, the dialogue is more instructional than plot-driven and functions to discipline the player. While this discipline will supposedly yield results for the player in improved game performance, this tutorial hardly represents the free agency often associated with interactivity. The *RK* game is structured so that the player interacts through a character from the film (Gandalf) with another character from the film (Gimli). In addition, the player must attend to the messages spoken by Gimli in order to play the game successfully. The subsequent successful game play, however, is often rewarded with more forms of promotion, as I shall demonstrate.

Beating the Levels

Once the player has beaten the tutorials, the character upgrade screen is revealed. Depending on how many combination moves the player has used during game play and how many enemies have been vanquished in that level of play, various numbers of points are awarded upon completion of the level. For example, *TT* scores a player's attacks as "perfect," "excellent," "good," or "fair," and points are rewarded accordingly. When these points have been tallied, the avatar's "special abilities" become available; each character has increasing levels of special characteristics that can be bought using the points awarded in previous levels. Thus, the better one plays, the more complex characters one can create, and one can earn special moves, more power, or more strength. This feature makes the game much more involved than simply beating a level. The ability to manipulate, however minutely, a character's attributes gives a much more interactive feel to the game. In addition, after beating certain levels, the player is given a reward: special features, or Easter eggs, are unlocked that give the player access to interviews with actors from the *LOTR* films or to other promotional materials.

For example, after beating the tutorial in *RK*, the player is rewarded with a feature entitled "Film Concept Art." This short, animated sequence shows artwork that was eventually incorporated into the film's design: drawings and paintings imagined from the book that evolved into the film. Other, similar features that can be unlocked during the game include "Game Concept Art" and "Film Production Stills." Some of the concept art shown does not appear to be from the game at all, and most of the film production stills do not represent scenes depicted in the video game. At the end of the "Film Concept Art" gallery, however, the player is presented with an ad for the Houghton Mifflin Harcourt book *The Art of the Return of the King*. In other words, part of the reward for playing the tutorial is information about another ancillary product from the *LOTR* franchise.

In addition to the "Film Concept Art" gallery, the player who beats the tutorial also gains access to the "Hobbits on Gaming" feature, a brief interview with three of the actors who played the main Hobbits in the films, who talk about playing *RK*. Dominic Monaghan, Elijah Wood, and Billy Boyd are all shown playing the game and commenting on each other's gaming abilities; Wood is supposedly the best. Although the short refers to them as Hobbits, the actors appear as themselves, so the audience must rely on intertextual knowledge of the *LOTR* films in order to recognize the actors as Hobbits. In addition, the visual representation of actors playing the video game of a film in which they have appeared directly relates the experience of the game to the experience of the film. Therefore, the second reward for successfully completing the *RK* tutorial is the suggestion that playing a character in the film is similar to playing the same character in the video game. Indeed, while the various still galleries provide opportunities to pitch other products, it is these interviews that seem to invite the player into the experience of the franchise, with the offer to "play the movie."

In *TT*, a similar Easter egg is opened after the player beats the second level. In this segment, Wood is interviewed about his impressions of the video game, and he talks about his own interest in the game and how he likes to play it all the time (he is also shown playing and reacting to the game). He notes how much the character of Frodo in the game looks like him: "Isn't that cool? It's so cool!" Wood discusses his voice-over work for the game, comparing it to his acting in the film. Most important, he identifies himself as a fan of video games and refers to the *TT* game as being a "geek bonus." Indeed, by identifying himself in this manner, he also identifies with the game player, who may be reluctant to identify as a geek but is likely willing to identify with Wood as a game player.

RK also contains an interview with Wood, which is unlocked after the "Shelob's Lair" level is completed. "Shelob's Lair" begins with footage from a scene in the film in which Sam must rescue Frodo by battling a giant spider. In the interview that follows, Wood bemoans the fact that, in the film, "Frodo never gets to join in the battle," and he rejoices that, in the game, "I'm a playable character now." He talks about the enjoyment he experiences in playing the game: "It is cool to take my character from the film and be able to implement it in the game; it is pretty awesome." Like many of the other features on *RK*, this interview is labeled with the character's name—this particular feature is labeled "Frodo"—but shows the actor, Wood, talking about his character. The line between character and actor is blurred when actors such as Wood refer to the characters as "themselves" and claim they get to "play themselves" in the game. In this case, Wood seems to suggest that he found playing the character of Frodo in the game more enjoyable than playing the character in the movie. In other words, for winning this level, the player is rewarded with the suggestion that playing the game is not just as good as being in the film: it is better.

By way of contrast, beating the second level of *TT* opens an interview with Ian McKellen (Gandalf), who claims he has never played video games. He further distances himself from the game player by admitting that he does not have the skills to play video games. McKellen talks about his voice-over work and claims that the Gandalf whom the player encounters in the game is close to the character he portrayed in the film. He also acknowledges that, in spite of his unfamiliarity with video games generally, the experience offered in *TT* is more immediate than the film experience.

RK also contains an interview with McKellen, who this time claims to be "reporting" from the offices of EA Games about the production of the game. During this interview, he observes, "[I]t's a curiosity of Gandalf that there are many of us." He refers to himself as an actor playing Gandalf, the stuntmen playing Gandalf, the digitized Gandalf of the film, and now, "Gandalf in the game." This, of course, implies that, just as McKellen played Gandalf in the film, the player can also play Gandalf in the game and thereby step into the role. In other words, the player can now be counted among the multitude of Gandalfs. McKellen makes the comparison even clearer by the end of the interview: "if you can't be in the movie, you might as well play the game. It's the next best thing." This operates as a strong endorsement of the game. McKellen identifies himself not as a fan of video games but as a film actor, yet he admits that a medium with which he is unfamiliar has the ability to make good on its offer to deliver the experience of the film.

This offer is made even more overtly when the player wins the second level of *RK*, "Escape from Osgiliath." After a cut scene that uses live-action scenes from both *RK* and *TT*, the game unlocks the extra feature "Sean Astin Interview," during which the actor, who plays Sam in the film, reveals that "the game is sort of an action/war game [and] the aspect of Sam's character that comes through is the heroic/fighting side." The connection between the film and the game is further enhanced when Astin states that the character is "a realistic version of myself," and the game play looks "just like what we were filming on the set." In an interview available on *TT*, John Rhys-Davies gives a similar assessment of the game. The segment shows Rhys-Davies playing the game and remarking on how well his character has been depicted; he notes that 14 months of work went into the game so that it accurately reproduces the film. In this way, the player is again rewarded with a message that folds the experience of playing in the film into the act of playing the game.

Even the villains get in the game. After "The Road to Isengard" level is beaten on *RK*, the player unlocks an interview with Christopher Lee, who plays Saruman. Lee is shown recording voice-overs for the game, and he remarks that, as far as his character is concerned, "The power of Saruman lies in his voice." Lee's comment is interesting because it suggests that the actors' connection to their characters in the game, which are merely digital renderings, are almost as strong as their connection to their characters in the film. While Lee may not be physically present in the game (in the way that he is in the film), the player is told that the presence of Lee's voice is what really matters. Given that Saruman's power is one of the elements that propels the plot of *LOTR*, if his voice is the conduit of his power, then Lee's voice is more important than his physical presence.

In a feature that is unlocked at the end of the game, Andy Serkis, who plays the role of Gollum, is also interviewed. Like the characters in the video game, the character of Gollum is fully digitized in the film. This segment shows Serkis performing in a motion-capture suit, a full bodysuit that digitally registers movement, and this method of motion capture is often used in game and film production. This feature blends the elements of film and game production, showing that similar methods were used to generate the character of Gollum for both the game and the film. Serkis highlights this blending in the interview when he mentions that he did the voice-over work for the video game before he did the voice-overs for the film, suggesting that his work on the video game was not much different from his work on the film. It also blurs the temporal order of the film and the video game production, so that his work on the game seems to be the original and his work on the film, the copy.

A segment in *TT* on the making of the video game further blurs the distinctions between film and game production. In this segment, the lead game developer, Scott Evans, and his crew compare their tasks to the burden Jackson assumed in bringing these characters to life. For example, they discuss how props and entire scenes from the film were used to develop the CGI incorporated into the game. In addition, Evans and his crew talk about how they built the characters for the game, noting that the actors were very interested in helping. This claim, of course, is supported by the segments I have already mentioned in which the actors voice their interest in the games and their respect for the games' ability to capture the experience of the films. In turn, Evans and his crew repeat the claims made by the actors that the *TT* game expands on the experience of the film by elaborating the visual aspects of the film. Wood appears again in this segment to comment on how much the game resembles the film, a remark taken one step further by the game developers, who claim that the game *is* the film.

Indeed, although this segment is specific to *TT*, both games incorporate a great deal of live-action footage from the films, often providing cut scenes woven into the games in such a way that the games are connected to the films on a visual level, a narrative level, and at the level of actual game play. For example, the final level of the *RK* game—"The Crack of Doom"—features Frodo in a battle with Gollum over the Ring. In this respect, it mirrors the climactic scene in the film, during which Frodo and Gollum fight over the top of a volcano for the power of the Ring. When the level has been completed, the voice-over by Gandalf tells of the end of this age of Middle Earth while the video displays live-action scenes from the end of the film. The last level of *TT* also reflects the concluding scene of the film, with Gandalf arriving at the last minute to save the day.

When the games are finished, the player has the option of playing them again and choosing a new avatar based on additional characters from the film.[27] After completing *TT,* the player can choose to play the game as Legolas Greenleaf, and successful play as this character opens up more Easter eggs with more promotional messages, for example, an interview with Orlando Bloom in which he marvels at the advanced technology used in the game. Two of the new characters unlocked in *RK* are Pippin and Merry, played by Boyd and Monaghan, both of whom are interviewed about the joy of playing as their characters in the video game. Monaghan even addresses the player directly, offering the persona of his character as a reward for beating the game: "this is now our treat to you as a gamer." This treat, and the reward for winning, is to play the game again. Although it is specific to *RK,* an interview with David Wenham, who plays Faramir, perhaps most succinctly articulates the synergistic purpose of both games: "The game itself has captured in all its aspects the excitement of the film *The Return of the King.*"

Interactivity and Incorporation

In these games, the player's interactivity is structured around a variety of promotional messages and synergistic practices. Design elements, sound and music, and most important, actual footage from the *LOTR* films are woven into the games. The tutorial for the games disciplines the player in relation to the reward structure, which allows the player to access special features upon beating the various levels. These features often contain messages that attempt to blur the distinctions between the experience of the films and the experience of the games. It is possible for a player to reject the synergistic force of these messages and to dismiss the "treat" of playing these games all over again. But such resistance would not be a product of the interaction of game play; on the contrary, it would be a rejection of the products of interactivity. Therefore, in the case of these games, interaction serves to incorporate the player into the media franchise. Successful players are rewarded with messages suggesting that game play connects them with the experience of Middle Earth as it is created in the film and assures them that they have made a wise media choice.

While it is clear that the game player is neither the passive subject of cinema spectatorship nor the ideologically resistant active television viewer, it should be noted that these approaches to reception practices were developed in contexts very different from the contemporary media environment. The twenty-first-century media consumer has a multitude of choices available from a variety of sources. To capture these consumers' attention, media producers must develop content that addresses a subject who is consuming products across a variety of media and delivery platforms. The *LOTR* franchise reveals how media conglomerates develop content that attracts audiences and motivates consumers. The video game is not just a metaphor for the media industry's strategies, as Marshall suggests; video game interactivity *is* a media industry strategy.

These games are inherently intertextual in relation to the *LOTR* franchise and seem to hail a subject who is not merely interested in playing the game, but who is part of the fan culture that surrounds the franchise. As Brian Ott and Cameron Walter note, "The intertextual allusions found in postmodern texts allow viewers to exercise specialized knowledge and to mark their membership in particular cultures."[28] In chapter 4, I will explore this strategic intertextuality further in an analysis of video games based on Marvel Studios' films. The games analyzed in this chapter demonstrate a strategic intertextuality that invites the player into the films and speaks to a subject who is part of a culture created by the *LOTR* franchise.

This incorporation reveals a paradox within the structure of these games. While they continually invite the player into the experience of playing the movie, the messages that offer this immersion actually interrupt and preclude any immersive game playing experience. For example, in *RK* (as in most games), the cut scenes forward the momentum of the plot. Yet, as Tong and Tan describe these scenes, "Once the cut scene commences, the player loses control of the character and therefore control of the camera. . . . [The cut scene] thus interrupts the degree of immersion that most . . . games seek to create."[29] In other words, the cut scene marks a point at which players are drawn out of the game and thus reminded that they are, indeed, playing a game. At first glance, this might appear to be a negative aspect of playing the game; for Tong and Tan, it is a necessary evil of the current video game industry.[30] However, for the franchise of *LOTR*, it may not be counterproductive at all. It is not in the interests of the film franchise producer to have any single product offer a completely immersive experience. For the cross-promotional and synergistic practices of a franchise to work, the consumer (and player) must be reminded that there are other products to be consumed.

What first appears paradoxical therefore makes sense from the perspective of the media franchise. The intertextual relationship of the media products reminds consumers of the other products in the franchise, inviting them to become part of the *LOTR* fan culture. For example, the cut scene introductions to and exits from the games serve a promotional function that subtly influences players to reexamine their own status as participants in the *LOTR* franchise in relation to the products available. While the games invite players to play the movie, those who benefit from the franchise want to remind players that there is not only a movie but also a DVD, a book, and other ancillaries. Therefore, the game experience cannot be complete, and needs interruption, to remind players of the other products to be bought and other choices to be made to further extend the *LOTR* experience.

Consequently, in the case of the *LOTR* franchise, incorporation is a practice not of immersion but rather of the continued consumption of a variety of media products. Given that video games figure prominently in this franchise, the interactivity associated with video game play needs to be reconsidered. Ideological resistance is not obtained in mere game play, particularly when that play stays within the parameters of the manufactured game. On the other hand, the corporate manipulation of this interactivity is not monolithic, and the *LOTR* franchise is a case in point. Although the franchise has been an incredible financial success, the apparent happy ending at the 76th Academy Awards proved to be illusory where Jackson and New Line were concerned. After winning the award, Jackson went on to direct and produce a remake of *King Kong* (2005) for Universal.

Although the film made $332 million in foreign markets, its domestic take was disappointing and exceeded its production budget by only $11 million.[31] Although Jackson has several more projects in the pipeline, *The Lovely Bones* (2009) is the only major motion picture he has released since *King Kong* premiered in 2005.

New Line fared much worse, and its success with *LOTR* was short-lived. The company's most successful film since *RK* was *Rush Hour 3* (2007), which brought in only $140 million domestically, exceeding its production budget by a mere $125,968. In addition, the studio was sued by Jackson over the ancillary rights revenue for the *LOTR* trilogy, and this litigation interrupted plans to move forward with the production of *The Hobbit*.[32] The suit was settled, Jackson took over as a producer, and the production is now moving forward with director Guillermo del Toro (*Pan's Labyrinth*, 2006; *Hellboy*, 2004).[33] *The Hobbit* and a sequel will be co-produced and distributed by New Line and MGM, and they are scheduled to be produced simultaneously in the manner of the *LOTR* trilogy.

The fact that New Line is not unilaterally expanding on this franchise speaks volumes about its status post-*RK*. Although New Line was owned by Time Warner, it previously operated independently; this is no longer the case. After New Line was sued a second time (this time, by the Tolkien estate), the studio was stripped of its independence and incorporated into Warner Studios, and the parent company basically took over New Line's production roster.[34] Although the studio appeared to do everything right where the *LOTR* franchise was concerned, and the video games certainly document a well-planned synergistic strategy, it also had some major failures on its hands. For example, when New Line released *The Golden Compass* (2007), the film grossed only $70 million domestically, while it had a production budget of $180 million.

The *LOTR* games illustrate how interactivity is not inherently liberatory where the player is concerned. Synergistic tactics are apparent in these video games, and these tactics have been structured into the game play. In spite of New Line's problems, the appeal of Middle Earth will be tapped once again with *The Hobbit* films. At this point, there is no official announcement about video game spin-offs for either film, but should New Line or Time Warner (and MGM) decide to release games, they most likely will be designed with synergy in mind.

COPPOLA SLEEPS WITH THE FISHES

When EA Games released *The Godfather* video game, it symbolized a new benchmark in collaboration between the film and video game industries. As mentioned in chapter 1, game spin-offs are often reserved for a certain type of film, and these films are not always noted for their artistic merit. There are exceptions: *The Lord of the Rings: The Return of the King,* for example, won several Oscars. But *The Godfather* is more than just an Oscar-winning film. The American Film Institute ranked it third, just behind *Citizen Kane* (1941) and *Casablanca* (1942), among the greatest American movies of all time.[1] In other words, this video game release is a little different from those accompanying films such as *300*. *The Godfather* game was released almost 35 years after the film appeared in theaters, and in that time the film had become part of the cinematic canon. The game is also remarkable because EA was able to secure the vocal talents of James Caan, Robert Duvall, and even Marlon Brando to recreate their characters from the film. Brando only did a day of recording, and only a portion was usable, but his voice-over work occurred just a few months prior to his death, making *The Godfather* video game one of the last performances of this iconic stage and film actor.[2]

Al Pacino did not participate in the making of the video game, and neither did the director, Francis Ford Coppola. A year before the video game's release, Coppola complained that he was never even asked to participate in the project:

> They never asked me if I thought it was a good idea. I went and I took a look at what it was. . . . What they do is they use the characters everyone knows and they hire those actors to be there and only to introduce very minor characters. And then for the next hour they shoot and kill each other. . . . I had absolutely nothing to do with the game and I disapprove. I think it's a misuse of the film.[3]

Some of those involved in the game's development dispute Coppola's claim that he was never consulted, but ultimately the video game was produced without his contribution. His absence is significant because his name has been so closely associated with the film. Indeed, when film critics describe the director's role as that of an "auteur," Coppola is often given as an example, with *The Godfather* offered as evidence of his artistry.

In film studies, artistry and agency have primarily been attributed to the director in what is known as *auteur theory*. Originally developed by French film critics, auteur theory invests the director with the agency of authorship and maintains that the qualities that distinguish certain directors (auteurs) can be identified in the cinematic text. As an academic theory, auteurism fell out of fashion years ago when it was supplanted by psychoanalytic and ideological criticism. Still, the concept of the director-as-author has survived in the popular consciousness as well as in the industrial practices of film production and promotion.[4] For this reason, identifying a film with a bankable name, such as Steven Spielberg or Martin Scorsese, becomes an effective way to market the film and attract an audience. Where *The Godfather* game is concerned, however, Coppola's imprimatur was not needed.

EA Games released *The Godfather* video game in March 2006 on the PS2, Xbox, and PC platforms. The game sold well and was listed at the top of several sales charts. Tor Thorsen reported that, by May, the game had sold over a million copies; extrapolating from the retail price, *The Godfather* video game would have generated about $40 million at this point.[5] In addition, the game was favorably reviewed, albeit with some reservations about the ability of the game to convey the experience of the film.[6] GameSpot, however, indicated that consumers rated the game even more highly than did reviewers.[7] After the game's initial release, EA further exploited *The Godfather* franchise by marketing new versions of the game for Sony's PS3, Microsoft's Xbox 360, and Nintendo's Wii. In April 2009, *The Godfather II* video game was released, including an online version for the Xbox 360.

Despite Coppola's lack of involvement, *The Godfather* was successfully repurposed for the video game market. Although the developers at EA indicated to *Variety* that they were disappointed by Coppola's decision not to participate in the development of the game, his participation was ultimately unnecessary: "'But we're not trying to create the movie,' said David DeMartini, exec [*sic*] producer of the game. 'We're taking it to a new medium with new consumers where we hope we can do something great like he did and [author Mario] Puzo did.'"[8] DeMartini viewed *The Godfather* video game as a media product distinct from the film and from Coppola's and Puzo's authorship.

In this chapter, I will consider the issue of authorship as it pertains to *The Godfather* video game. I begin with a discussion of auteur theory generally and Coppola's authorship specifically. I will then discuss how video game theories have conceptualized player agency in relation to narrative construction and authorship. My analysis of *The Godfather* game will look at how the player's agency contributes to *The Godfather* narrative. I conclude that Coppola's authorship is replaced by the agency of the player, and this erasure serves the interests of both EA and Paramount Studios.

Coppola, the Auteur

When auteur theory was first introduced in the film journal *Cahiers du Cinéma* in the 1950s, the proponents of the theory argued that a cinematic text was the artistic expression of its director and thus the director should be considered to be the author of the cinematic text. Influenced by French postwar politics, these critics positioned directors as agents in an antagonistic relationship with the studio system, who were struggling to realize their vision against the alienating machinery of capitalistic enterprise. Andrew Sarris is credited with importing auteur theory for the study of American film, and in doing so he emphasized romanticism by revering those directors who succeeded in their struggles to have their visions realized.[9]

The stories that have been told about *The Godfather*, and the stories that Coppola has told himself, often present him in the romantic terms of the auteur. As Michael Schumacher notes, when Paramount Studios acquired the rights to Puzo's book, they offered the project to several directors, who all refused, before settling on Coppola.[10] Once hired, Coppola began butting heads with both Paramount's president, Stanley Jaffe, and its head of production, Robert Evans. First, Coppola fought to hire Marlon Brando to play Vito Corleone, and then he fought to cast Al Pacino, who at the time was not well known, to play Michael Corleone. In addition, he insisted on setting the film in New York in the 1940s, although the studio wanted a contemporary film shot in either Kansas City or St. Louis to hold down production costs. Other conflicts with the studio would arise related to the production schedule and budget overruns.

Yet beyond these various conflicts lie complex issues of authorship. After all, Puzo's book was successful in its own right, and he collaborated with Coppola on the screenplay for the film. Yet Schumacher claims that this collaboration was not equitable, for "[a]s Puzo learned, a director has the ultimate creative control, and while the two were generally compatible, there were inevitable disagreements."[11] In an interview in *Film Comment,* Coppola leaves little doubt as to who won those disagreements:

> I wrote the *Godfather* script. I did the adaptation. I credit Mario completely with creating the characters and the story. On the other hand, his book took in a lot more than what the film took in. I feel that I took in the right parts. I also did a lot of things in that movie that people thought were in the book that weren't. The act of adaptation is when you can lie or when you can do something that wasn't in the original but is so much like the original that it should have been.[12]

It is clear that Coppola imagines he created a story that, while related, is distinct from Puzo's. Not as clear, but certainly implied by Coppola, is his belief that his story is the better one.

Jeffrey Chown outlines the differences between Coppola's film and Puzo's book, noting Coppola's additions as well as the parts of the book he chose to exclude.[13] For example, in the book the Johnny Fontane character, a singer with acting aspirations, is much more prominent than in the film. Another difference Chown identifies is Coppola's portrayal of morality: "Rather than seeing the Mafia as a metaphor for how individuals cope in an unjust world as Puzo does, Coppola sees it more as a metaphor for the predatory, selfish, and ruthless aspects of modern contemporary America."[14]

In spite of Coppola's example, auteur theory has had many detractors. Some critics of auteur theory argue that the theory does not account for the inherently collaborative process of filmmaking, while others maintain that the romantic notion of the struggling director fails to capture the realities of contemporary film production. Since the decline of the studio system, directors not only operate with more autonomy, they often enjoy significant career benefits through the support of the film industry.[15] As Virginia Wright Wexman observes, "[T]he Hollywood industry has become increasingly concerned with marketing directors as salable commodities."[16] It is not that film production has ceased to be a collaborative process. On the contrary, in spite of the decline of the studio system, there is more corporate oversight of the creative process than ever, but it still often serves the interests of film studios to represent the director as an authorial agent.[17]

As a result, the studio practice of promoting films has often gone hand-in-hand with promoting directors, with beneficial results for both directors and studios. Coppola is a case in point. As opposed to the traditional view of the auteur laboring under the constraints of the studio system, Chown has argued that Coppola problematizes this director-studio relationship, because he has worked with various studios in order to establish himself as a respected and widely recognized director. According to Chown, Coppola has carved out a career by taking major studio jobs in order to support his work with American Zoetrope, the studio he founded with George Lucas.[18] And Schumacher points out that Zoetrope was in financial trouble when Coppola agreed to direct *The Godfather*.[19] Chown observes that Coppola was able to make his mark on the film's text through the traditional means of directorial agency: cinematic stylization, mise-en-scène, and narrative additions. Through these means, Coppola achieved critical acclaim for the film, and *The Godfather* became a significant text in his cinematic oeuvre and put Coppola on the map as a bankable Hollywood director.

Paramount has also benefited from *The Godfather* franchise, and the three films have earned about $450 million worldwide.[20] In addition, this trilogy of films has won a total of nine Oscars in addition to several other awards. Whatever problems Coppola may have caused Paramount, the studio appears to have been quite happy to call on his services for both the second and third installments in the trilogy and to put Coppola's name above the title in both films. Indeed, it seems that Coppola's artistry was highlighted in Paramount's marketing strategy for the film franchise. As the studio now attempts to revive and extend the franchise through the video game, however, it has decided to move forward without Coppola. If Coppola's work on *The Godfather* signifies the romantic figure of the auteur, then Paramount's use of the franchise signifies how authorship is something that studios are willing to manipulate and exploit, assign and reassign.

Interaction and Authorship

One of the reasons that traditional auteur theory fell out of favor is because it promoted a singular concept of artistry rooted in a centered subject, and it could not accommodate the broader approaches to authorship that began to emerge. For example, Roland Barthes challenged the authority of the author to signify the meaning of texts, arguing that readers also have agency in bringing meaning to texts.[21] In other words, readers can use texts to construct experiences apart from those imagined by authors. This critique of authorship influenced the emergence of the cultural studies approach, which considers how consumers use popular culture for their own ends.[22] This consumer-centered critique of authorship can also be found in the critical study of video games, with many scholars maintaining that the interactive qualities of games give the player authorial agency. For example, Ben Sawyer, Alex Dunne, and Tor Berg claim that "the notion of interactivity means that the decisions and skills of the player will move the story in a certain direction."[23] Henry Jenkins provides a concrete example of this agency when he discusses the *Matrix* franchise.[24] In addition to the three films, there was a video game containing narrative elements that linked all of the texts together. The video game bridged the narratives of the second and third films in the trilogy, and by successfully completing the game the player became an agent in the storytelling process.

Authorship has been a problematic concept in game studies. Some scholars argue that players enjoy the agency of authorship, and they characterize interactivity as allowing the game player to construct narratives.[25] Yet, as Aphra Kerr mentions, the concept of the auteur is difficult to apply to video games, because video game production is a collaborative process that undermines the concept of single authorship.[26] Eric Zimmerman suggests that the association between the film and game industries may have inhibited new thinking about narrative and game play: "The commercial game industry is suffering from a peculiar case of cinema envy at the moment, trying to recreate the pleasures of another media. What would a game-story be like that wouldn't be so beholden to preexisting linear media? Good question."[27]

It is a good question, and it raises another: if the video game industry has chosen to produce games that reference the narrative structure of older media, is it wise to develop theories that ignore those conditions? I agree with Zimmerman that the potential of video games as a medium may not be realized if we continually look to traditional ideas of narrative and authorship. Yet "cinema envy," or at least the practice of repurposing narrative films into video games, requires a critical examination of these traditional ideas of authorship. Video game production is a collaborative process, but so is film production. And it is important to remember that the director's artistic agency is sometimes a conceit of the film industry.[28] Indeed, agency is invested in the director by the film studio, and Coppola is a case in point. Therefore, the agency of the game player, and the interactivity that is offered as a means of authorship, may also be regarded as products of the game developer.

The Godfather game provides a unique opportunity to critically engage questions of game players' agency and authorship, because the film it uses as its reference point is so closely associated with Coppola as auteur. I do not intend to posit the game player as an actual creative author, nor am I recreating the game player in the romantic terms of the auteur. When I mention the game player's "authorship," I refer to the ability of the player to move the game narrative forward, but I imagine this agency to be part of the conditions of production. As I have mentioned, the film industry locates authorial agency behind the camera and invests it in the person of the director, because it serves the interests of the film industry. I maintain that the video game industry, in order to create a marketable experience, serves its own interests by locating authorial agency behind the control pod and investing it in the person of the game player.

The Godfather game illustrates this shift in agency. As I mentioned earlier, DeMartini noted that *The Godfather* game was intended to introduce the film franchise to a fresh new audience. After all, those born when the first two *Godfather* films were released are now well into their 30s, and even those born when the last film was released in 1990 are now entering their 20s. While it is uncertain whether this new audience is respectful, or even aware, of Coppola's reputation as an auteur, they are likely to be familiar with the video game experience, and the game's producers are banking on this familiarity. In order to reach this audience, the producers of the game have created an experience in which the game player assumes agency in constructing a character and advancing a narrative.

In the analysis that follows, I will demonstrate how *The Godfather* video game player participates in the completion and expansion of the film's narrative. While the game is structured around the narrative of *The Godfather* film, the game player elaborates on that narrative, completing action that is not depicted in the film but is nevertheless important to the narrative of the film.[29] In this game experience, Coppola's agency becomes immaterial, because the story the game player completes stands alongside, but apart from, the one told by Coppola.

An Author You Can't Refuse

The Godfather game begins with the Paramount logo on the screen while the theme music from the film plays underneath. The opening credits roll, framed by the famous puppeteer logo that has been used to market and promote *The Godfather* franchise. This logo has been used on the cover of the book and on the one-sheets for the films, but it serves a rather unique function in the game. As is common in video games, certain non-playable characters (NPCs) appear in the game, and the player must interact with these characters in order to either advance the plot or obtain information about the next level or mission. These NPCs are often indicated by an icon that hovers over their heads, and in *The Godfather* game, that icon is the puppeteer logo. While this logo has generally signified the franchise, within the context of the game it prompts the player to advance the game narrative. *The Godfather* game is an open world that allows players a great deal of latitude in advancing the narrative. Therefore, players can determine when they want to interact with these designated characters and learn the information they have to offer.

Don Vito Corleone speaks at the beginning of the game: "Some day, and that day may never come, I may call upon you to do a service for me. Until that day, consider this a gift." This line of dialogue appears early in the film, when the don agrees to avenge an assault on the daughter of Amerigo Bonasera, the undertaker. Apart from this similarity, the beginning of the game differs significantly from the film. The game opens on a scene set in Little Italy in 1936. Johnny Trapani is paying his respects to Don Corleone, and then he meets his wife, Saraphina, on the street to make plans for the evening. A nearby storefront explodes, and Saraphina is concerned for the safety of their son, who was playing in a nearby alley. Johnny runs into the alley looking for his son, and there he is confronted by members of the Barzini family. A fight ensues, which serves as a tutorial for the combat in the game, and after the tutorial concludes, a cut scene shows Johnny being gunned down by the Barzinis. Johnny's son attempts to run to his dying father, but he is stopped by the don, who makes him look away. Don Corleone tells Johnny's son to restrain his anger until he can seek revenge.

The game then jumps forward in time. Johnny's son is now an adult, and this character serves as the avatar for the game. At this point, the game player is allowed to exercise agency in creating the central character of the game narrative. Specifically, the player can alter the avatar's appearance and give him a name. Most games allow the player to set options related to costume and equipment, but in this game even the facial features of the avatar can be carefully morphed to the player's taste. The choices seem endless: the hairline, hair color, facial hair, eye shape, brow angle, nose shape, body size, and lip size can all be chosen by the game player.

The creation of the avatar has been identified as an important aspect of the game player's agency.[30] While many scholars have argued that the avatar provides an opportunity for the expression of personal identity, *The Godfather* game reveals the possibility of a different sort of agency. In addition to creating an identity that the player can assume, these choices allow the game player to define the physicality of a character in a manner parallel to the exposition of a novel or the casting of a film. In spite of the many choices available, the player cannot make the character a different race, as one might in some role-playing games, nor change the gender of the avatar. Still, the player can name the character and thereby define its identity. (The default name given to the avatar is Aldo, and that is the name I will use throughout this analysis.)

The practice of styling the avatar is not limited to defining facial features. As the game play advances, the player can choose to change Aldo's style of dress as well. These changes are more than cosmetic, because they also enhance Aldo's characteristics; specifically, as Aldo dons better clothes, his respect points increase. Again, similar choices are found in other games, and in the context of *The Godfather* game this agency allows the game player to determine how this character will develop through the progression of the game narrative. In other words, the game player creates and develops a character that resides outside of the narrative created by Coppola. However, as I will show, this character becomes central to Coppola's narrative.

The Godfather game presents an open and expansive world. Though limited to the terrain of New York, its boroughs, and New Jersey, within this terrain there are numerous exteriors and interiors. Although the player advances through the various missions of the game following a chronological progression, the player can decide when to begin a mission and how many "side-quests" to complete between missions. These side-quests do not necessarily advance the plot of the game in a linear sense, but they do allow the character to acquire attributes and equipment that make the game play easier as the difficulty of the missions increases. In *The Godfather*, these side-quests include extorting money from businesses, hijacking cargo trucks, bribing officials, and carrying out hits on the enemies of the Corleones. These missions also require the game player to direct Aldo to various locations in the game world. Therefore, the player is at liberty to proceed to either a side-quest or a mission and to move about in the world of the game according to these choices. Through these choices, the game player determines how and at what pace the game narrative progresses.

As the game player directs Aldo through the game, he becomes a member of the Corleone family by completing various missions that are related to the film's narrative. These missions, however, are never actually represented in the film; instead, Aldo completes actions that take place off-camera. In this manner, the game player adds to the narrative of the film by playing out and completing actions that, while central to the advancement of the film's narrative, do not actually appear in the film. These specific missions begin once the player has Aldo interact with one of the NPCs mentioned earlier, specifically the ones identified with the puppeteer logo. Therefore, the logo often signifies points of narrative intersection, where the player moves into the narrative as it is defined by the film.

For example, Aldo's introduction into the Corleone family plays out as part of the film's opening scenes. At the beginning of the film, Don Corleone is celebrating the marriage of his daughter, and one of the characters mentions that the don must honor the requests made of him on this special occasion. In the film, scenes of the wedding are interspersed with scenes showing the don receiving guests and listening to their requests. In the game, after the tutorial and after the game player has chosen Aldo's characteristics, a CGI cut scene depicts the Corleone wedding.

The game, however, shows a request that does not appear in the film: Aldo's mother, Saraphina Trapani, asks the don to intervene on her son's behalf. Saraphina tells the don that Aldo has fallen in with some "bad men," and he needs the don's help. This scene is sandwiched between CGI reproductions of two scenes that do appear in the film: one of Luca Brasi rehearsing his congratulatory speech for the don, and another of Brasi delivering the speech and wishing that the don's daughter will bear him a grandchild who is a "masculine child." It is after this second scene that the don asks Brasi to locate Aldo, which he does. These cut scenes lead into another tutorial in which Brasi trains Aldo as a fighter and welcomes him into the Corleone clan. In this way, the action from the film becomes interwoven with the action of the video game, and Aldo's story becomes intertwined with the narrative of the film.

The opening scenes of the film also provide the context for Aldo's first mission. When the undertaker Bonasera asks the don to exact justice for his daughter who has been violently sexually assaulted, the film never depicts how this justice is carried out. It is understood, however, that the don calls in the favor when later in the film he asks

Bonasera to do him a service and make his son's (Sonny Corleone's) bullet-ridden body presentable for his funeral. In the game, this justice *is* depicted, and Aldo assists the characters of Paulie Gatto and Monk Malone in hunting down and beating the men who attacked Bonasera's daughter.

The mission in the game is called "A Grave Situation," and in it Aldo must first locate Paulie and Monk, after which the three of them chase the two men who are their targets into a graveyard, where they commence to brutally assault them. One is beaten with a baseball bat, while the other is thrown into an open grave and hit in the face with a shovel. The action is bloody and violent, as is most of the action in the game, but when justice is served the mission is complete. Therefore, the character of Aldo, in service to the family, completes a mission that becomes important to the actual action of the film, and the game player fills in the narrative of the film by completing the necessary off-screen action. By completing this mission, Aldo fulfills a favor for the family, one that becomes important to the don later in the film.

Throughout the game, the successful completion of the missions opens up clips from the film. These clips not only reward the successful completion of the missions, but they are also interwoven with game play and help to set up the narrative contexts for Aldo's subsequent missions. One of the more infamous scenes from the film is included in these clips, and it involves the family's efforts to secure a movie role for the singer Johnny Fontane. In the film, Tom Hagen (the unofficially adopted son of Vito Corleone) meets twice with Jack Woltz, the director who has refused to cast Johnny in the movie. Hagen's efforts to persuade Woltz to change his mind fail both times. In the second meeting, however, Woltz is more cordial to Hagen and takes him on a tour of his stables, where he shows off his prized racehorse. After refusing Hagen a second time, Woltz wakes up to find the decapitated head of his prized horse in his bed. In the game, however, the player creates the action that the film does not depict. Aldo helps Rocco Lampone to sneak into the stable, then keeps a lookout while Rocco butchers the horse; in a rare instance of tasteful discretion, this action takes place off-camera in the game as well. Aldo then follows Rocco into the mansion, throttling security guards along the way, and keeps watch while Rocco sneaks the head into Woltz's bedroom. The next cut scene is a CGI reproduction of the scene from the film in which Woltz discovers the head, and his screams are heard from outside the mansion. In this manner, the game player completes the action responsible for one of the most memorable scenes from the film, and Aldo once again helps the family to accomplish their goals.

As Aldo succeeds in his efforts to help the Corleones, he is rewarded by promotion within the family. As the game advances, the player unlocks film clips that depict the family meeting with the drug dealer Virgil Sollozzo and that show Vito's youngest son, Michael, being briefed for his hit on Sollozzo and Captain McCluskey. After these clips, Aldo is assigned another mission ("A Recipe for Revenge") by Hagen to plant a gun in the bathroom of Louis's Restaurant in preparation for Michael's hit. Aldo drives Michael to the restaurant, then sneaks into the back of the restaurant to hide the gun. A cut scene follows, showing Aldo listening in on the conversation among Michael, Sollozzo, and McCluskey. After Michael completes the hit, Aldo must drive him to the docks in preparation for Michael's escape to Sicily. Again, Aldo completes action that is never depicted in the film, yet is essential to the film's narrative. In the film, Michael is told that the gun will be hidden in the bathroom of the restaurant and he finds it there, but how it is planted and by whom are never revealed. Aldo not only is responsible for planting the gun, but he also makes sure that Michael arrives safely at the dock for his escape to Sicily, a trip that sets in motion several other important narrative developments in the film, including Michael's first marriage.

The significance of this mission can be understood in relation to the film's narrative, because Aldo's actions are instrumental to a pivotal point in the film. In this scene, Michael is introduced into the violence of his family's business, violence that earlier in the film he claimed to reject. In addition to their significance to the film, Aldo's actions are also significant to the advancement of the game. After he completes this mission, he meets with the don and is invited into the family and promoted: Aldo is now officially a Corleone, and this promotion also improves the avatar's abilities.

Aldo's inauguration into the family carries with it additional responsibilities, and he now must exact revenge for the family. After film clips show Connie Corleone Rizzi being beaten by her husband, Carlo Rizzi, and Sonny attacking Carlo for beating his sister, Aldo is required to complete a mission called "Change of Plans." In this mission, the player guides Aldo as he follows Sonny's car to the Little Italy Toll Plaza. Once they arrive at the plaza, the game presents a cut scene that reproduces an important scene from the film, in which Sonny is gunned down by members of the Stracci and Tattaglia families. The CGI of the game captures the same images found in the film of Sonny's body flinching and jerking under the seemingly endless hail of bullets.

After the hit, Aldo follows the assassins' car to a warehouse and factory. Once there, Aldo finds and interrogates the toll booth attendant, who tells him that the order for Sonny's hit came from the Tunnel Club. At this point, the player has the option of letting the attendant live or executing him with a bullet through the head. If the player chooses the latter option, the action earns more respect for Aldo (after all, he is exacting revenge for the family). Aldo then goes to the Tunnel Club and shoots his way through the club until he comes upon the Tattaglia underboss. Here, Aldo must interrogate the underboss, yet the underboss resists Aldo's efforts to gain information, that is, until the player directs Aldo to threaten a young woman who is in the room. Only then, in order to protect the woman, does the underboss relent and disclose that the Tattaglias were not the only family involved in Sonny's death. The player then directs Aldo back to the Corleone compound, which triggers a cut scene showing a CGI reproduction of a scene from the film in which the don meets with the heads of the various families in order to establish peace. Aldo becomes an important catalyst for this meeting, because he is responsible for uncovering the information regarding Sonny's death and the involvement of the other families.

In one of the last missions, "Baptism of Fire," Aldo meets Michael in a graveyard adjacent to a church. Michael asks Aldo to meet Peter Clemenza at a flower shop, setting in motion a series of hits that Aldo must complete. Specifically, Aldo must execute the four dons of the Stracci, Cuneo, Tattaglia, and Barzini families. In the film, these hits are depicted as the various executions that punctuate the baptism of Michael's godson. The scene is significant in the context of the film because it represents Michael's ultimate descent into the violence that he once eschewed. In the game, it is Aldo who must carry out Michael's violent ambitions by going to the various locations and completing the executions himself. Therefore, the game player is responsible for directing Aldo through the action that brings *The Godfather* film to a close. Upon completion of these missions, Aldo receives additional respect and is promoted by Michael to the rank of underboss. At this point, the film's narrative ends, but the video game's narrative goes on. Aldo is given the task of eliminating all of the rival families by infiltrating and bombing their compounds. After he destroys all of the family compounds, Aldo goes to the Corleone compound for an important meeting. A cut scene shows Aldo being promoted to be the godfather; he stares into the camera as he is pronounced the don.

What happens to Michael at this point in the video game is unclear. In the closing scene of the film, Michael assumes the duties of the don. In the subsequent film, *The Godfather: Part II,* Michael has moved to Lake Tahoe and from there oversees operations in Las Vegas, Miami, and Havana. According to *The Godfather II,* the Corleones' interests in New York are handled by another character, Frank Pentangeli. In *The Godfather* game, however, Pentangeli does not appear, and Aldo, a character created by the game player, replaces him and thereby supplants the narrative arc between the first and second films created by Coppola. Where the game is concerned, it is Aldo, not Pentangeli, who inherits the New York businesses.

Once the player completes all of the missions—including 15 contract hits and 84 business extortions—and seizes 56 rackets, 8 warehouses, and 4 transportation hubs, Aldo is promoted one last time, and the game is successfully completed. In the final cut scene, Aldo looks out the windows of his office, high above the New York skyline. One of his underbosses comes forward to kiss his hand, and Aldo takes a seat at the head of a conference table as he assumes the position of the don of New York City.

■ ■ ■ Clearly, *The Godfather* game will never replace the film, nor will it achieve the movie's iconic status in American culture. The game is more like a by-product of the film's status. Nevertheless, the game contains its own narrative, and it is up to the game player to complete that narrative. This game narrative not only intersects the narrative of the film, but it does so at crucial moments, and the game player is responsible for completing actions that never appear in the film yet are essential to it. Almost all of these actions require the use of violence, sometimes extreme violence.

On this point, Coppola's dismissal of the game is accurate: it does depict characters that appear to do little but shoot and kill people.

It is not surprising that Coppola would characterize the game as a misuse of the film, because its graphic depiction of violence is something he tried to contain in the film. Coppola did not want to depict the same level of violence found in Puzo's book, and he completely distanced himself from the actual Mafia members who served as consultants for some of the actors in the film. Indeed, it was Coppola's idea to temper the final acts of violence in the film with scenes from the christening.[31] It is as if Coppola wanted to tell the story of the American Mafia while maintaining his artistic distance from the subject, so as to remain the auteur.

The original novel was a work of fiction, Coppola's film depicted that fiction, and the game is just another fiction—another way to tell the story. The story told by the game player, however, does not enjoy Coppola's artistic distance. Instead, the game player must complete the often violent actions of the story that Coppola kept off the screen. The game player must do the dirty work: sneaking into stables, crawling around a toilet, and actually pulling the trigger several times. The story told in the game is the one Coppola was reluctant to tell, but the game player cannot afford to be so artistically detached. In this way, the game finds legitimate footing in *The Godfather* franchise; it tells the rest of the story, the part that is closer to the street and to the action.

EA Games might have preferred to have Coppola's cooperation in the development of the video game; after all, Coppola was approached. But Coppola's refusal was ultimately immaterial. As DeMartini points out, the game is a media creation quite distinct from the film.[32] Indeed, the game creates an experience in which the player assumes the agency of authorship and elaborates on the narrative of the franchise. In the case of *The Godfather* video game, the game player creates an experience and completes a narrative that, while related to the film, is ultimately distinct from Coppola's own artistry because it embraces the violence that he eschewed.

Nevertheless, while the game player is playing, it is another hand that holds the puppeteer strings depicted in the logo. Any agency the player enjoys is granted by Paramount, which holds *The Godfather* franchise. The decision to move forward with the game without Coppola's input allowed Paramount to assert real authority over *The Godfather.* For the producers of the game to reach a new audience and to create a player-centered experience, Coppola's reputation was unnecessary. Indeed, if the game is any indication, Coppola's agency as an auteur has been replaced by the game player's own agency. In other words, the video game gave Paramount the opportunity to remove Coppola from *The Godfather* franchise, and the game does facilitate that removal. After all, Coppola was not telling the entire story.

Film is and always has been a collaborative practice, and the agency of the auteur is a construct that has been maintained because it serves the interests of production. Perhaps more to the point, the film industry creates conditions that enable the auteur's sense of agency. In a similar way, the interests of production are well served when the game player is invested with creative agency, because this agency makes the experience of games more interactive and potentially more pleasurable. Therefore, just as the film industry uses the construct of the auteur to stroke directors' egos and to market films, the video game industry provides a similar sense of creative agency to engage the game player. In both cases, this sense of agency is conferred by the media producers and is constrained by their interests.

Coppola's absence from *The Godfather* video game may be an indication of the future status of the auteur in the film industry. The ideas of cinema as an art form and of the director as an artist still survive in the art and independent film markets. While the auteur may remain important to the production and marketing of smaller films, this conceit may have a diminished role in the larger blockbuster films that drive the box office and inspire video games. For example, there are probably few people who would rush to see a Gore Verbinski film simply because it was a Gore Verbinski film. But people did just that, and in record numbers, when *Pirates of the Caribbean: Dead Man's Chest* became the highest-grossing movie of 2006. As the film and video game industries continue to collaborate, the auteur may become less important as a marketing tool and may even become a distraction from the key agency that must be marketed: that of the game player. Finally, independent film production, the bastion of the auteur, is in sharp decline. Some production houses have folded, while others are facing significant financial problems.[33] If these trends continue, not only Coppola, but the romantic figure of the auteur itself, may be sleeping with the fishes.

FOUR

MARVEL GOES TO THE MOVIES

In the fall of 1995, Isaac Perlmutter and Avi Arad made what can best be described as a desperate play to save Marvel Comics. Perlmutter and Arad were partners in Toy Biz, a company that manufactured—as the name clearly suggests—toys. More specifically, Toy Biz produced licensed products, and part of its line was devoted to Marvel superhero action figures. A couple of years earlier, Perlmutter had become disenchanted with the licensing fees that Toy Biz received and wanted to secure intellectual property for his company. In 1993, he approached Ron Perelman, one of the most notorious and powerful men on Wall Street, who at the time owned controlling interest in Marvel. The two men cut a deal in which Toy Biz would retain the rights to all of Marvel's characters in perpetuity, while Perelman would receive a 46 percent share of Toy Biz.[1] Two years later, that deal was about to turn toxic for Toy Biz. Marvel was on the verge of bankruptcy, and there was a possibility that the company would discontinue publishing. If that happened, Toy Biz would hold the licensing rights to characters who had no future. Perlmutter and Arad offered to Perelman what they thought would be the solution to Marvel's problems: make movies.

Ten years later, Marvel was not only making movies, but also setting box office records. Yet, to get to that point, Marvel had to change hands several times and weather some high-level financial wrangling. Some observers might never have imagined that a comic book publisher would be the subject of Wall Street machinations, but sometimes in the world of finance the unimaginable happens. The wheeling and dealing that surrounded Marvel Comics' ownership has been meticulously chronicled by Dan Raviv in *Comic Wars: Marvel's Battle for Survival.*[2] The battle Raviv describes lasted for about 10 years and involved numerous banks and countless lawyers—and all over a bunch of comic book characters, one of whom has green skin and wears purple shorts.

Marvel Comics was founded in 1939, but for many years it remained in the shadow of its main rival, DC Comics. Marvel, however, started to hit its stride in 1961 when Stan Lee and Jack Kirby began developing the superhero characters that would come to symbolize the Marvel brand: *Spider-Man,* the *X-Men, Iron Man,* and the *Fantastic Four.* Spurred by the success of these characters, Marvel began to diversify in the 1960s and 70s; the Marvel Entertainment Group was developed, and its comic book characters began appearing in TV movies and animation, as well as on a variety of merchandise.[3] In the 1980s, Marvel became more actively involved in animation production and was purchased by New World Entertainment. Three years later, New World sold the company to Perelman, the chair of MacAndrews and Forbes, who then gained controlling interest in the Marvel Entertainment Group.

Perelman is a well-known Wall Street power player who is notorious for hostile takeovers, due in part to his highly visible and hostile acquisition of the Revlon company. He is also noted for his involvement in the junk bond market and his association with the equally notorious Michael Milken. Indeed, as Raviv explains, the junk bond market had a great deal to do with Perelman's interest in Marvel:

> The true wizardry in these arrangements was that the bond issuers also protected themselves by setting up "holding companies," separate legal entities to carry the somewhat uncertain obligation to pay off the bonds. Perelman issued junk bonds, for instance, through companies with names such as "Marvel Holdings" and "Marvel Parent Holdings."[4]

Because of these dealings, Marvel did not do well under Perelman. By 1995, the company's stock had lost half its value, and Perelman soon threw it into bankruptcy. In addition, in 1996 one-third of Marvel's New York publishing office staff were let go, and it appeared that publication of Marvel comic books would cease.

Meanwhile, another noted corporate raider, Carl Icahn, smelled the blood in the water and began buying up the bonds issued by the Marvel holding companies created by Perelman. Perelman, who had an ongoing rivalry with Icahn, was prepared to put up a fight and he did, resulting in several legal proceedings. In the end, Icahn emerged with the controlling interest in Marvel and went on to become the company's chair in 1997. The struggle did not end there, however. After Icahn made a power grab for Toy Biz, Perlmutter and Arad began battling him for control of Marvel. When Marvel finally emerged from bankruptcy in 1998, after much legal wrangling, Icahn was out and Perlmutter and Arad were running Marvel. They were now at liberty to pursue the idea they had presented to Perelman three years earlier: to produce movies based on Marvel's characters.[5]

Although Marvel did not hold the film rights for all of its characters, the bankruptcy allowed the company to nullify previous deals, and soon Perlmutter and Arad were able to secure contracts with major film studios. If box office returns are any indication—and in Hollywood, they are—their plan worked quite well, because Marvel Studios has produced some very successful films. All three of the *Spider-Man* movies are among the top 20 in all-time domestic box office receipts.[6] The three *X-Men* movies have generated over $600 million in gross revenues, and *Iron Man* was the second highest grossing film of 2008.[7] Marvel has had missteps (*Hulk*, 2003; *Daredevil*, 2003), but its success is undeniable.

Marvel Studios' early efforts were co-production deals with major studios such as Fox and Columbia, in which these studios paid Marvel a licensing fee for its characters and Marvel received 50 percent of the merchandising revenue from the films.[8] With the success of its earlier films, however, Marvel Studios now has enough capital to produce films independently. *Iron Man* was its first independent production, which Paramount distributed for 10 percent of the gross receipts. Following the success of that film, Marvel negotiated a five-picture distribution deal with Paramount at the rather low rate of 8 percent.[9] Under this deal, Marvel assumed the production costs for its movies and was poised to reap the benefits at the box office.

The annual report Marvel issued for 2008 painted an interesting picture of the company: net sales from film production generated $254 million, while net sales from publishing only generated $125 million. These figures are actually not all that surprising, and they support Matthew McAllister's observation that Marvel is more concerned with product licensing than with comic book publishing.[10] Where the bottom line is concerned, Marvel is looking less and less like a comic book publisher; for stockholders, that may be good news, but for comic book fans, it may not.

As Henry Jenkins has argued, fan cultures are often developed from the ground up by the fans themselves, who are actively involved in the cultures they develop around films, television shows, and comic books.[11] They often police the boundaries of these cultures and are on the lookout for the inauthentic. Fans can thus be highly suspicious of the corporations that produce popular culture products, and highly contemptuous of products that appear to be solely motivated by profit. Given the company's history and its near destruction at the hands of corporate raiders, the fans of Marvel's comics had reason to be suspicious, particularly when Marvel began teaming up with other corporations to produce films based on the comic book characters. Therefore, Marvel had to appease its fan base while still producing films that would have a broader appeal.

In this chapter, I will demonstrate how the video games released for Marvel's films addressed that challenge. I begin with a brief discussion of comic book fan culture in general and the culture created by Marvel specifically. I then will discuss the practices of strategic intertextuality and how these strategies are augmented by the use of new media and video games in particular. Finally, I will provide a textual analysis of several of the games released in tandem with some of Marvel's most successful films.

Marvel Fan Culture

In *Comic Book Culture,* Matthew J. Pustz analyzes a variety of comic book fan cultures.[12] As Pustz notes, there are several different types of fans who engage comic books with differing levels of commitment. One of the marks of a true fan is the amount of knowledge that person has about a particular comic book character and narrative. Pustz explains, "[C]omic book fans enjoy being experts, even when there is no one with whom to share the knowledge. When . . . fans are fortunate enough to be among the like-minded, an element of competition also certainly exists."[13] Indeed, anyone who has spent time in a comic book store has probably overheard the conversations that take place in these spaces, in which the customers (and often the store clerks) debate the relative merits and powers of various superheroes and the minutiae of plot points. This observation is not meant to suggest condescension toward those who participate in fan culture, because I too am a comic book fan.

As I prepared to write this chapter, I dug out my old Marvel comic books. My favorite superheroes were the Fantastic Four (FF); I collected their comics starting with issue #73 in 1968 and ending with #119 in 1971. (I never really got over Jack Kirby's departure from the series.) At that point in time and in that location (Wichita, Kansas), there were no comic book stores like the ones that exist today; I bought my comic books at drugstores. As Pustz notes, however, Marvel was successful at building a fan culture from the top down. The editors at Marvel used the comics to speak to fans directly in the "Bullpen Bulletin" section that appeared in the back of the books, often opposite ads for "spy-scopes" and "sneezing powder." These sections were clearly designed to promote Marvel's comics and included a list of newly published issues and their plot lines. For example, the "Bullpen Bulletin" in FF #119 lists "*Hulk* #148: At last! Our mountainous man-monster finds Jarelia—the girl from the emerald atom! And—Rick Jones conquers the universe! The wildest one yet!" The back section also contained "Stan Lee's Soapbox," in which Lee addressed various issues related to Marvel Comics, such as why the price of the books increased from 15 cents to 20 cents (this was 1971, after all).

On the "Fantastic Four Fan Page," Marvel even articulated its own hierarchy of fans with the "Hallowed Ranks of Marveldom," a list of titles that denoted a fan's commitment. Pustz hones in on the purpose of these rankings when he argues that status was often tied to consumption of Marvel products. For that reason, I never rose above the level of RFO (real frantic one), bestowed on those who purchased "at least 3 Marvel mags a month." Through the "Bullpen" and the fan pages, Marvel built a sense of community around Marvel comics, creating its own fan culture. As Pustz observes, "Lee and his company were selling more than just comic books. They were selling a participatory world for readers, a way of life for its true believers."[14]

In 1995, however, that way of life seemed to be coming to an end. Not only was Marvel teetering on the brink of financial collapse, the quality of the comics themselves was also in decline, due to Perelman's decision to significantly cut the publishing staff. Fans responded with a boycott of Marvel Comics in 1996, initiated by Bob Kunz, a student at Duke University. Although the boycott was primarily motivated by changes in the Marvel characters (specifically, changing Spider-Man's secret identity from Peter Parker to Ben Reilly), the fans were also aware of what was happening on the corporate level. As Raviv notes, "The fans knew America's comic book empire—its output translated and exported to over a dozen countries—had become a mere cog in the conglomerate machine that lumped it in with a perfume maker, a camping supplies manufacturer, and a toy company."[15] After about a year, the boycott ended, Peter Parker returned, and although the company was still in the midst of a financial and legal battle, Marvel continued to publish.

Obviously, Marvel fans had legitimate reasons to be suspicious of corporations and profit motives. Consequently, Marvel had to be careful when it began teaming up with major studios to produce its films. After all, if fans were skittish about the involvement of global conglomerates, then Marvel's decision to collaborate with Sony and Fox/News Corporation could easily give the wrong impression. Marvel realized that it needed to be mindful of its fans and produce films that respected their characters' comic book origins, so the company would not appear to be motivated merely by profit. The films and ancillary products had to be designed so as to reflect and respect the characters that Marvel fans held so dear.

Intertextual Strategies

As I noted in previous chapters, video games based on films often contain extra features that give the player additional information about the film and its relationship to the video game. I have already discussed these features in relation to interactivity. Another way of theorizing these features draws on the work of John Fiske and his distinction among primary texts (actual content), secondary texts (promotional materials), and tertiary texts (fan-produced materials).[16] Although Fiske was writing about television culture, these categories lend more general insight into the relationship between content and the extra materials found on ancillary products. The categories work more generally because the types of relationships that the film and television industries try to establish with their audiences have become increasingly similar, due in part, no doubt, to the synergistic practices that permeate large media conglomerates.

Fiske uses the term *intertextuality* to describe the relationships among primary, secondary, and tertiary texts. Yet, as Brian Ott and Cameron Walter point out, there are two types of intertextuality: that which is produced through the interpretation of media and that which is generated in its production.[17] They label this latter form "strategic intertextuality" because it can involve intertextual references that are designed to appeal to a particular audience. For example, *The Simpsons* television show is highly self-referential, containing allusions to its various episodes and even to the tertiary texts that have been produced on fan web pages. This use of intertextuality, Ott and Walter argue, creates specialized knowledge around particular media products, and these references invite further consumption of media content. In other words, it is difficult to understand the reference to a previous episode of *The Simpsons* unless one has watched that particular episode, and it is difficult to understand the references to the fan culture unless one is part of that culture. In this way, we can see the producers of media forming fan culture from the top down.

The references to *The Simpsons*' fans' web pages also reveal the importance of digital media in the evolution of intertextual relationships. The internet has provided fertile ground for the production of tertiary texts, evident in the large numbers of fan sites and blogs. However, there are also highly visible examples of how this digital terrain has been infiltrated by corporate interests. *The Blair Witch Project* site is a prime example of how the web can be used to generate buzz about a film project: it parades studio promotional messages as special knowledge about the "real story" behind the film. More recently, *Snakes on a Plane* used blogs to generate interest in the film, and it is reported that the project was put back into production in order to shoot additional footage to incorporate suggestions that were made on these blogs. This corporate use of the supposedly independent blogosphere has been duly noted on the *Pajiba* media blog. Publisher Dustin Rowles writes:

> The problem, of course, is that somehow the very folks the blogging world was set up to rail against have gradually manipulated their way into our daily discourse. It's incremental, to be sure,

> but the collective "we" is slowly being absorbed into the corporate agenda. . . . And what does this have to do with *Snakes on a Plane*? Well, everything actually. Because were it not for the blogging world, *Snakes* would've been dumped unceremoniously into a crowded weekend box office, where it would've netted around $5 million before silently making way for a *Dukes of Hazzard* sequel.[18]

I would argue that the co-option Rowles describes is a necessary reaction of the film industry to the rapid expansion of digital media, particularly given how this proliferation has increased market fragmentation. Film studios realize that they must address fans in order to attract fans, and to do that they must appeal to fans' specialized knowledge. In addition, studios have discovered that, by using digital media, they can package information that appeals to fans into ancillary products and thereby harness that appeal to a revenue stream.[19]

Anna Everett has argued that the digital media have created "new interactive protocols, aesthetic features, transmedia interfaces and end-user subject positions," and she coined the term *digitextuality* to refer to these developments.[20] Perhaps the most apparent change is the interactive immediacy that digital media have brought to the intertextual experience. For example, the use of hypertext allows the web surfer to click on a word or phrase and go immediately to another web page that elaborates on that word or phrase. With video games, the experience is not immediate in the sense of clicking on a link (although game interfaces often resemble web interfaces), but it is immediate in the sense of the immersive experience that games provide. In other words, these intertextual relationships are embedded in the game experience and are often part of the reward system of the game. This immersive immediacy creates a strong rhetorical advantage over traditional media. While intertextual relationships are often spread over a variety of different media, including these relationships in a single product—like a video game—greatly increases the likelihood that they will be consumed, particularly when they are offered as a reward for successful game play.

My analysis will look at several games that have been released in conjunction with the *X-Men, Fantastic Four, Spider-Man,* and *Iron Man* films.[21] I see these games as strategically intertextual, linking the films with their comic book origins. My analysis will focus on both the game play and the special features of these games, and I will argue that the games attempt to appeal to the special knowledge held by fans of the Marvel characters. In this way, the games not only connect the films with the comic books, they also reassure fans that Marvel has not abandoned them. In other words, the games speak to fans in a manner similar to the "Bullpen Bulletin," cultivating the fan culture through digital means.

The Fantastic Four

First published in 1961, the FF comics introduced characters Reed Richards; his college roommate Ben Grimm; Richards's girlfriend, Sue Storm; and Sue's brother Johnny Storm. Richards is a scientist who takes the other three with him on a space flight to the Van Allen belt, where their spaceship is bombarded by cosmic radiation. These cosmic rays transform the four humans, endowing them with superpowers. Richards becomes Mr. Fantastic, with the ability to stretch and contort his body in seemingly endless ways. Grimm's skin mutates into orange rocks, giving him superhuman strength, and he adopts the name the Thing. Sue Storm becomes the Invisible Girl (later renamed the Invisible Woman), a name that captures her ability to disappear and produce unseen force fields. Finally, Johnny Storm becomes the Human Torch, with the abilities to ignite his body and to fly.

The film reproduces this original story, albeit with some narrative changes. In the film, Richards collaborates with Victor von Doom in the space experiment, and von Doom is also transformed by radiation into Dr. Doom, a villain who appears later in the comic book series. The movie follows Richards (Iaon Gruffudd), Sue Storm (Jessica Alba), Grimm (Michael Chiklis), Johnny Storm (Chris Evans), and Dr. Doom (Julian McMahon) as they discover their new powers, and it concludes with a final battle between the FF and Dr. Doom.

The FF is neither the best- nor the worst-performing franchise for Marvel. The film took in over $150 million at the box office, outperforming both *Hulk* and *Daredevil,* but it did not come close to the success of the *Spider-Man* films. In terms of its ancillaries, again the performance was strong, but not as strong as Marvel's most popular properties. In fact, the company's annual report noted that the revenues generated by the FF franchise were below what was expected.[22]

Indeed, a lot had been expected from the FF. Arad predicted that the film franchise would be Marvel's biggest, and he negotiated with Fox to move the film to a summer release to maximize the opportunity to market toys and the DVD.[23] Although it did not fulfill its promise, the FF nevertheless represents Marvel's state-of-the-art marketing strategy for ancillary products. Just as important (to me), they were my favorite characters. I therefore will begin my analysis with the *Fantastic Four* video game.

The game begins, as do all of Marvel's films and video games, with the Marvel logo. This logo shows the pages of comic books being rapidly flipped, so the viewer sees several Marvel scenes and images. The images then morph into the word "Marvel," suggesting that this new film studio is essentially about its comic books. The first cut scene of the game is a CGI scene of the characters approaching Dr. Doom, who comes to life and recites a line from the film: "This ends now." Grimm then provides a narrative that brings the action of the video game in line with the actual narrative of the film. He also introduces the members of the FF, who are represented as video game characters with the appropriate character screens.

The next cut scene deviates from the visual action of the film, although it stays true to the narrative. In the film, Grimm is sent outside the space station to perform repairs when the cosmic storm hits. His work is not depicted in the film, but in the game the player gets to guide Grimm through the repairs. After the player completes this level, there is a cut scene that reproduces the actual scene from the film in which the FF are transformed by cosmic radiation.

Another important scene from the film that is reproduced in the game involves a traffic pileup on the Brooklyn Bridge. This game level begins with a cut scene of Grimm walking across the bridge when he witnesses an accident with a fire truck. The player must save the truck in the same manner that Grimm saves it in the film. After Grimm saves the fire truck Sue, Johnny and Reed arrive at the bridge, and the scene serves as the precipitating event that officially forms the FF. Indeed, after this level of the game, the FF is formed in a manner that reflects the action of the film, and the player gets to perform the action that completes this pivotal scene.

These are all examples of intertextual references to the films, and are similar to the references that are made in the other games I have examined. This game, however, also includes intertextual references to the comic books, including characters and scenes that have appeared in the *Fantastic Four* comic books, although they do not appear in the film. For example, Annihilus is a villain who appears in both the video game and the comic book; he is also represented in the original comic book cover art that is included as a special feature in the game. Another example of this intertextual strategy occurs in the level in which the FF battle the Mole Man. The Mole Man is a villain who appeared in the first issue of *Fantastic Four,* and the game play on this level culminates in a fight that resembles a scene depicted on the cover of the first comic book. After the various members of the FF fight the Mole Man and his minions in subways and caves, the action is taken up to the streets for a final confrontation. In this scene, the player directs all of the members of the FF in a battle with the Mole Man and his "mole crusher"; the visual tableau of this fight is similar to the one depicted on the comic's first cover more than 40 years ago. Again, the original comic book cover is included in the video game to remind the player of the close connection to the source material.

In the *Fantastic Four* comic books, the arch-nemesis has always been Dr. Doom, who is wisely included both as the villain of the film and an adversary in the final "boss" battle (a major battle that concludes a level and/or the game) in the video game. In the concluding levels of the game, the FF battle Dr. Doom several times. After the first battle concludes, the player views another CGI cut scene that resembles the action from the movie. Grimm then fights Doom, and after Grimm wins the battle, the scene shifts to a final fight with a transformed Doom, where again the game play reproduces action from the film. The final fight ends when Richards subdues Doom by using water to solidify him while he is in a molten metal form; this conclusion mirrors the climactic scene from the film as well. The subsequent cut scene shows the characters walking down the street, agreeing that the four of them combined are greater than each one individually. Johnny lights the sky with the *FF* logo as Doom is being shipped to Latvia. The control panels of Doom's containment device flicker ominously, and in this way the video game concludes in the same manner as the movie, with the hint that a sequel is imminent.

After the player beats the various levels, special features are unlocked in the video game. These include concept art from the game, covers from the original comic books (including the original covers depicting the Mole Man's crusher and Annihilus), and interviews that show the actors and writers discussing the film and the video game. For example, Chiklis begins his interview by discussing how he was a fan of the FF from the time he was a child; he reveals that he told his brother of his ambition to play the character of Ben Grimm. He also comments on the quality of the video game and remarks that he can imagine playing the game with his daughters and that in doing so he will finally be considered "cool" by his kids. Gruffudd talks about how he replicated his character in the film through his voice work and how the fans will enjoy the thrill of being able to play scenes that appeared in the film; he is then shown recording voice-overs in a studio in tandem with scenes from the game. Evans also recommends the gaming experience by suggesting that players will be able to "relive the movie experience and go beyond by playing all of the Fantastic Four characters as superheroes."

In another interview, writer Zak Penn discloses that he has been involved in other Marvel film projects and has worked with Activision, the video game's producer. He states, "[S]ince I was intimately involved with the *Fantastic Four* on the studio production side, and since I am kind of an avid gamer anyway, they asked me if I would help coordinate the writing of the script for the video game." In other words, Penn identifies himself as someone who is familiar with the franchise and the activity of game play, thus suggesting that he was the ideal choice for constructing the narrative of the video game.

Writer Marty Signore explains that he and Penn worked together on several previous projects, but their collaboration has been personal as well as professional. As Signore notes, "[T]he thing that Zak and I did most together was play video games." Consequently, Penn invited Signore to become involved in the *FF* video game project. Signore goes on to observe that, as the film and video game industries become more closely associated, there are more opportunities for screenwriters to work on video game content. This observation exposes the synergistic connections between these two industries, but at the same time it is a testament to the fidelity of this video game to the film. While the same creative talent did not work on both the video game and the film, the same type of creative talent did. In other words, screenwriters have become game writers, and therefore the production of a cinematic narrative and a game narrative have become similar practices.

The game for the second film, *Fantastic Four: Rise of the Silver Surfer* (2007), offers the same intertextual strategies as the first game but is a poorly produced sequel. As I played this particular game with one of my students (it allowed for multiple players), we could not help but remark on its poor quality.[24] The game only loosely follows the plot of the film, and the Silver Surfer character does not appear until late in the game play. The likenesses of the actors are used for the avatars, but none of the actors provided the voice talent. The extra features are much more limited than those found in the first game. After beating the game, the player can choose between costumes that reflect how the FF looked in the comic books of the 1960s, the 1980s, and the 1990s. Successful game play also unlocks five original comic book covers, and there is a CGI blooper reel of Grimm that plays while the ending game credits roll, similar to the blooper reels employed in other CGI-animated feature films. In spite of this game's weaknesses, the two games taken together demonstrate two intertextual strategies: the first links the games to the films, while the second links the games to the comic books. These same strategies can be found in other games based on Marvel's films.

Spider-Man

The character of Spider-Man first appeared in 1962 in the pages of Marvel's *Amazing Fantasy* #15, and he was given his own comic in 1963 with *The Amazing Spider-Man* #1. In the comic books, Peter Parker is a 15-year-old boy who receives his powers through a bite from a radioactive spider. Peter is being raised by his Aunt May and Uncle Ben, and it is Uncle Ben's murder at the hand of a robber that convinces Peter to use his powers to become Spider-Man. The film adaptation, *Spider-Man* (2002), brought in $821 million in worldwide box office receipts, an amount surpassed by *Spider-Man 3* (2007), which grossed $890 million. *Spider-Man* 2 (2004) was the weakest performer of the trilogy, if $783 million can be described as a "weak" box office performance.[25]

The game for the first *Spider-Man* film opens with a CGI trailer that abstractly references the original story of Spider-Man: a spider is shown biting Peter Parker. This scene segues into various shots of the Green Goblin, and it is clear that the villain of the film will also serve as the villain in the game. After the opening, there is a cut scene in which actor Tobey Maguire, speaking as Peter Parker, provides some brief exposition, describing the murder of his uncle and his own responsibility in the murder. This scene provides a CGI reproduction of Spider-Man's fight with the wrestler Bone Saw McGraw and the subsequent robbery of the wrestling promoter. Therefore, the game begins at a point that comes much later in the film's narrative, but at a pivotal point for the development of the Spider-Man character.

After these cut scenes, the player is given the choice of playing the tutorial or beginning the game. It is a rather false choice, because the opening mission of the game provides tutorial prompts. If the player chooses the tutorial, however, Peter (Tobey Maguire) provides comments on the acquisition of skills. As he scales a building, Peter says, "This is incredible. I can't believe I'm doing this." Actually, it is the player "doing this," so Peter's experience in discovering his superpowers is mapped onto the player's experience.

The first mission begins with Peter looking for the Skulls gang, a member of which robbed the wrestling promoter. He then discovers that it was this gang member who also murdered his uncle. Peter must web-sling through the buildings of New York and battle gang members until he finds the warehouse in which the murderer is hiding. In important ways, the game still parallels the film, because the player's effort to master the practice of web-slinging between buildings mirrors Peter's first tentative efforts in the film. In addition, the character of Spider-Man is rendered in the same type of homemade costume that Peter first dons in the film when he fights Bone Saw.

The game differs from the film in that it presents the murderer as part of a well-populated gang, while in the film he appears to act alone. Still, the final act of this mission allows Peter to avenge his uncle's death, which references an important plot point in the film. After Peter subdues the murderer, a cut scene follows in which Peter realizes that he could have stopped the villain *and* saved his uncle. The murderer then backs up and falls through the warehouse window, as he does in the film. Another cut scene begins with a voice-over from Uncle Ben telling Peter that he is becoming a man and must assume important responsibilities. Peter then gives a great deal of exposition, introducing the characters of Mary Jane Watson, Peter's romantic interest; Harry Osborn, Peter's friend; and Harry's father, Norman.

At this point in the game, the player is introduced to the character of Norman Osborn, and a cut scene shows Osborn's transformation into the Green Goblin. The player will progress through several missions before confronting the Goblin in the final battle. Along the way, however, the player must also confront several other villains who do not appear in the film, including the Shocker, the Vulture, Kraven the Hunter, and Scorpion. All of these villains have appeared in the *Spider-Man* comic books; for example, the Vulture appeared in the second issue of the comic and Kraven in the fifteenth. In the game, these characters are woven into a narrative that also reflects the narrative of the film. For example, after Spider-Man defeats some robots produced by Oscorp (Osborn's company), a cut scene follows in which he is attacked by the Vulture, then almost run over by a van driven by the Shocker. As players complete the missions and defeat these villains, they also contribute to the production of a text that links the film with its comic book origins. Indeed, as in *The Godfather* game, in which the player fills in the story told in the film, this game reveals the rest of the Spider-Man story, one only partially told in the film.

The strategy guide for the game further augments the links between the films and the comic books. The back of the guide includes a section called "Making the Game," which contains much of the concept art for the game design and the storyboards for the cut scenes and the game play. Storyboarding is a common practice for the planning of live-action films, and storyboards often resemble comic books: they contain sequenced illustrations of scenes for a film, and the illustrations are broken into boxes that suggest progressive action. The storyboards, which are used to help the cast and crew of a film imagine what the scenes will look like, draw on the same imaginative capacity that comic book readers use to link the action in one box to the action in the following box. Thus, the strategy guide reveals in a more general sense the connection between films and comic books as visual media.

In addition, the "Making the Game" section contains interviews with several of the people who worked on the video game. Some of these interviews contain the same type of cross-promotional statements found in the *LOTR* games. For example, in describing his role, Chris Soares, the creative director for the game, observes, "I directed the team in creating a game that shared the same vision as Sam Raimi, the film's director."[26] In these interviews, the game crew is asked to list their favorite films, games, and hobbies (unsurprisingly, many list playing video games). They are also asked which additional villains from the *Spider-Man* comics they would have liked to include in the game. Here, the game designers display their familiarity with the comic books. For example, the audio director, Sergio A. Bustamante II, chooses Black Cat: "Is she a villain? Ah, I don't care. . . . I would have liked to see her in our game."[27] Bustamante's uncertainty about Black Cat actually reveals some familiarity with the character, because in the comic books she was alternately a friend and a foe of Spider-Man. Bustamante subsequently saw his wish fulfilled when the video game for *Spider-Man 2* was produced. Other wishes were fulfilled as well, as the most popular answer to the question about missing villains was Venom, who later appeared in both the *Spider-Man 3* film and video game.

The *Spider-Man 2* video game begins with opening credits listing the voice talent of Maguire, Kirsten Dunst (as Mary Jane), and Alfred Molina, who plays Otto Octavius (Doctor Octopus or Doc Ock), the villain of the film and the main villain of the game. The game begins with Maguire providing the narrative context in a voice-over in which he explains that his is a "boy meets girl" story, as the camera swoops through the cityscape and finds Spider-Man sitting on the ledge of a building. A tutorial follows, and after it is completed a cut scene shows Peter arriving late for class and being admonished by Dr. Connors for his poor performance as a student. This scene captures the theme of the film, which is Peter's inability to balance his two identities and their separate responsibilities. Indeed, the film focuses on the conflicts Peter encounters in fulfilling his duties as Spider-Man while trying to pursue a romantic relationship with Mary Jane. Thus, this game, like the first one, reflects several elements of the film. Also, as in the first game, several characters appear in the *Spider-Man 2* game who do not appear in the film. Foremost among them is Black Cat, who introduces some conflicts of her own.

In the *Spider-Man* comic books, Black Cat proves to be a complex character as she is both a romantic interest and an enemy. In the game, Spider-Man first encounters Black Cat after he battles thieves in an art gallery. Spider-Man follows her through the city, and when he catches up with her, she asks him if he always chases girls "who brush him off." Later in the game, Spider-Man must chase Black Cat again because he suspects that she has stolen some jewelry. After he catches her, a cut scene plays in which Spider-Man demands that Black Cat return the jewels. She introduces herself as "Black Cat, and I just crossed your path." Afterward, Spider-Man hurries to Mary Jane's place, only to find a note telling him to meet her at the movies. He then has to zip to the theater in a short period of time, all the while mulling over his encounter with Black Cat.

In this way, the character of Black Cat serves as the "other woman" and further complicates Peter's already complex relationship with Mary Jane. In this instance, a character from the comic books is used to highlight the dramatic tension that exists in the film. For example, in the film Mary Jane develops a relationship with John Jameson, to whom she becomes engaged. In the game, once Peter learns of Jameson's proposal, Black Cat shows up and asks why Spider-Man would want a "boring girl" with a "boring life." She invites Spider-Man to go on a mission, and together they battle the Shocker and his minions. When the engagement is announced, Peter asks Mary Jane why she decided to marry Jameson, and she explains that she has chosen him because "he is there." Peter then contemplates Mary Jane's marriage and views himself as "just waving from the sidelines." He then meets Black Cat, who talks about why her life is so much fun and tries to get him to understand that he needs to enjoy himself.

Black Cat then solicits Peter's help to deal with an incident at a warehouse. A cut scene shows the sale of some robotic weapons, and then Black Cat and Spider-Man must battle these robots. After the battle, Spider-Man contemplates whether he should abandon his identity as Peter Parker, but he concludes that he must maintain both identities and keep them separate. He then tells Black Cat that he will not see her any more; he is not like her and needs the balance of a normal life. Black Cat tells him rather ominously that once she crosses his path, she does not go away, but she also tells him that he should go see Mary Jane. Afterward, Peter goes to Mary Jane's apartment and tells her that he knows he wants to be with her; she replies that she is getting married to John and that she "can't do this." Peter concludes that he was out of his mind to pursue a relationship with Mary Jane, and he needs to keep away from her.

Of course, he cannot, because the final battle of the game involves saving Mary Jane from Doc Ock. This final battle mirrors the action of the film, including a fight that takes place on an elevated train. After Spider-Man finally defeats Doc Ock, Peter apologizes to Mary Jane and tells her that she should continue with her plans to marry Jameson. This is followed by a cut scene that shows Peter staring out his apartment window, until Mary Jane arrives to tell him that she wants to be with him in spite of everything. The game concludes with Peter's voice-over telling the player that he has a great burden, but Mary Jane makes it an easier burden to carry, and that he *is* Spider-Man. Completing the game unlocks a "movie theater," but the offerings in this special feature are not all that special: the player is able to view various demos and a credit roll of the game's production team.

It is important to note how the game introduces Black Cat into both the narrative and the action. The player, as Spider-Man, must interact with this character and even fight alongside her in some of the game missions. Therefore, she is a significant presence in the game, although she is absent from the actual film. Her presence in the game helps to connect the film to the comic books and, if Bustamante's comment is any indication, would be enjoyed by those familiar with those books. In any event, her flirtation with Spider-Man in the game reflects the relationship that is present in the comic books and provides a counterpoint to the romantic conflict that appears in the film.

Like *Fantastic Four: Rise of the Silver Surfer*, the *Spider-Man 3* game was a disappointment. As I took notes on this game, the student assisting me could not help but remark on its poor quality. For example, four consecutive missions are spent chasing and fighting the same villain, the Lizard, and defeating him in the final battle brings more a sense of relief that it is over than a feeling of accomplishment. In addition, it is clear that Kirsten Dunst did not participate in the game production—the Mary Jane game character does not look like Dunst, and the actress who did the voice work does not sound like Dunst. Finally, the game brings little new to the table except for the villains Sandman, Rhino, and of course Venom; Scorpion and Kraven are reused in the game. In other words, the *Spider-Man 3* game merely reproduces the intertextual strategies found in the other two games.

Iron Man

Marvel introduced the character of Iron Man in 1963 in *Tales of Suspense* #39. Iron Man is the alias of Anthony Stark, a talented engineer and industrialist and head of Stark Industries. As mentioned earlier, the *Iron Man* film was Marvel Studios' first independent production, and it has done quite well for the studio. In the film, Stark has just demonstrated a new weapon to the military in Afghanistan when he is abducted by terrorists.

The *Iron Man* video game stays fairly close to the film narrative. It begins with a cut scene of Tony Stark (Robert Downey Jr.) and Dr. Yinsen (another terrorist captive) working on the Iron Man suit that Stark will use to escape from the cave in which they are held captive. After the cut scene, the player directs Stark, using the suit to battle his captors. As in the film, Stark escapes from the cave, but Dr. Yinsen is not so lucky. The next playable mission involves Stark testing the improved Iron Man suit and trying it out on a flight. This mission is sandwiched between cut scenes in which the characters Obadiah Stane and Pepper Potts appear and discuss Stark's discovery that his company's weapons are being used by terrorists and his decision to discontinue manufacturing weapons at Stark Industries. This is also the pivotal plot point in the film, but the game diverges as Stark sets out to discover how Stark Industries weapons ended up in the hands of terrorists.

To locate these weapons, Stark must travel to several locations and battle many enemies who do not appear in the film. As in the other Marvel games, these characters are often taken from the comic books—examples include Whiplash and Titanium Man. As the player progresses through the game and Iron Man defeats these villains, different Iron Man suits are unlocked for the player to use. For example, after completing the "Flying Fortress" mission, the player can unlock the "Comic Book Tin Can" suit option. The text within the game notes, "The first-ever Iron Man suit was built during the character's initial appearance in 'Tale of Suspense, #39,' and this design originates from . . . *Iron Man* Vol. #4, #5." Although the suit looks primitive (resembling the Tin Man in *The Wizard of Oz*), it carries all of the functionality of the other suits. Therefore, the most important advantage it offers the player is a sense of nostalgia and the satisfaction that goes along with this special knowledge of Iron Man's origins.

Other than the suits, the game offers little in terms of rewards. The player can destroy weapon crates that are located throughout the game world, but these just unlock concept art slides. Trailers for the *Iron Man* movie are available at any time during the game and can be accessed from the "Bonus" menu in the opening interface. These include a feature entitled "*Iron Man* Unmasked," which is a series of interviews with people who worked on the game and with Iron Man co-creator Stan Lee and the editor in chief at Marvel Comics, Joe Quesada. In this segment, Lee is credited with changing comic books by introducing Iron Man, a superhero who is an antihero, a rich industrialist who is not imbued with special powers but constructs his suit out of necessity.

The segment is rather direct in its purpose, with Beejey Enriquez, an associate producer for Sega (the video game company), claiming that the game and the film will appeal to fans of the comic book. It is worth noting, however, that this goal is also achieved indirectly by the mere presence of the individuals in this special feature. Here we have the director of the film, Jon Favreau; the lead actor, Robert Downey Jr.; Stan Lee; the editor in chief of Marvel Comics; and several members of the game design crew all speaking to one purpose: support of the *Iron Man* franchise. (In fact, Quesada refers to Iron Man *as* a franchise.) These voices create a text that links the film to the comic books and positions the game as the central point of connection because this text resides on the video game.

The X-Men

The X-Men are a group of superhuman mutants led by Professor Xavier. The first *X-Men* comic book appeared in 1963. To date, there have been three *X-Men* films: *X-Men* (2000), which grossed $296 million worldwide; *X2: X-Men United* (2003), which grossed $407 million; and *X-Men: The Last Stand* (2006), which grossed $459 million.[28] Although Marvel has licensed many video games based on the X-Men, none are actual spin-offs of the films in that they do not really convey the action or plots of the films. There are two games, however, that both borrow heavily from the films and are clearly designed to extend the franchise. *X2: Wolverine's Revenge* was released in April 2003, about a month prior to the release of the film *X2*. The game was used to promote the release of *X2*, offering free tickets to the movie to those who purchased the game. *X-Men: The Official Game* was released in May 2006, about ten days before *X-Men: The Last Stand.*

Both of these games adhere to the standard practices of the video game spin-off, although *The Official Game* does so more closely. For example, *The Official Game* uses the likenesses and voices of Hugh Jackman, Alan Cumming, and Shawn Ashmore, and the player has missions that alternate the use of these actors' characters: Wolverine, Nightcrawler, and Iceman, respectively. The game also incorporates villains from the film, like Lady Deathstrike and Magneto. *Wolverine's Revenge* uses Jackman's likeness and even uses his visage on the box cover, but for some reason Mark Hamill, of *Star Wars* fame, provides the voice of Wolverine in the game. Consequently, Derek Johnson has argued that the game fails to provide a cohesive image of the Wolverine character.[29] In spite of this failure, however, the game gives Wolverine a primary position among the X-Men, and this primacy is part of Marvel's future plans for the franchise.

The Official Game loosely follows the plot line of the second film in the *X-Men* franchise, involving the character of Jason Stryker. *Wolverine's Revenge,* on the other hand, explores the character's past as a military experiment gone awry. Both games attempt to evoke elements of the comic books, albeit in very different ways. In *The Official Game,* the cut scenes are not the usual full-motion video, but limited animation scenes in which static images are moved across a background. The likenesses of actors are used, but in illustrated form, and the visual effect is similar to comic book illustrations.

To hammer the point home, these cut scenes are sometimes sandwiched between CGI renderings of comic book pages. *Wolverine's Revenge* takes a more traditional approach, allowing the player to unlock original comic book cover art through successful game play. Yet again, these games not only attempt to anchor the franchise in the actual comic books, but also connect the various parts of the franchise. *The Official Game* provides an explanation for Nightcrawler's noticeable absence from the third film: after the final battle in the game, a cut scene plays in which Nightcrawler announces that he is leaving the X-Men because he is not suited for that kind of life.

Also in anticipation of the third film, *The Official Game* contains a cut scene in which Jean Grey appears and seems to be manipulated into the violence that she will display in *X-Men: The Last Stand.* The concluding cut scene of the game is a static image of comic books that look as though they have been scattered on a table. The top comic book cover shows the characters of the film, with Jackson's face (as Wolverine) most prominent. The other covers, partially obscured by the top book, are actual covers from issues of the comic book, but the image clearly indicates how this franchise is going to progress and who will be its central character. The next film from the *X-Men* franchise was *X-Men Origins: Wolverine,* which Marvel released on May 1, 2009. As the title suggests, the film traces Wolverine's origin story, much like the *Wolverine's Revenge* game. The game thus functions as a precursor to the most recent film, but in addition to this game Marvel released the *X-Men Origins: Wolverine* video game on the same day as the film.

■ ■ ■ The video games that have been released for Marvel's films clearly contain strategic intertextual references. The games often reflect the narrative elements of the films, yet they also incorporate elements found in the comic books. By gesturing to both the films and the comic books, the games exist as texts that link the films to their comic book origins. This strategic intertextuality has the potential to appeal to Marvel's fans in two ways. First, the intertextual references draw on and add to the special knowledge the fans possess about these superheroes. Second, the references indicate to fans that Marvel, in spite of becoming a large media company whose financial interests in comic book publishing are now secondary, is still respectful of the characters and their comic book origins.

Although it is difficult to say how much fans take comfort in these references, the strategy is clear. One thing that is certain is that Marvel as a company is on much better footing than before, and there appears to be no immediate threat to its publishing interests. For a time, Marvel appeared to be a slow-motion train wreck, with Perelman at the throttle. Though Perelman had ambitions for the company and wanted it to become a "little Disney," Raviv notes that he was much more interested in acquiring companies than in running them.[30] The two men who pressed Perelman to make movies back in 1995, however, have realized his goal of turning Marvel into a little

Disney, and they now provide leadership for the company. Arad was in charge of Marvel Studios until 2006, when he left to form his own production company, and Perlmutter is now CEO.

As a point of comparison for Marvel's success, Disney has 4 films among the top 25 all-time box office hits; Marvel has 3.[31] Though Marvel lacks the history that Disney has in the film industry, as a company with a brand identity and characters that can move theater tickets and ancillary products, Marvel has become a lot like Disney, and in 2009 Disney actively sought to acquire Marvel as a wholly owned subsidiary. Although Marvel may generate more revenue in films than it does in publishing, the intertextual strategies in the video games suggest that the company understands that its comic books provide the substance of its brand identity. Indeed, these video games present Marvel as a company that is still very interested in comic books. If it becomes part of Disney, however, Marvel's future interest in comic books may go no deeper than a desire to turn those books into movies.

FIVE

DISNEY SAVES THE WORLD(S)

A couple of years ago, I was invited to spend Christmas in Tokyo with a friend who was a visiting professor at Dokkyo University. As we were planning my trip, she asked if there was anywhere I was particularly interested in visiting. There was: Akihabara. I was in the early stages of my research on video games, and I had become fascinated with the legend of Akihabara as a center for video game culture in Japan. She arranged for some of her Dokkyo students to meet us there and take us through the area, and they proved to be excellent guides.

Akihabara is an incredible, almost overwhelming experience. The area occupies only a few city blocks, but those blocks contain several multilevel video game stores, some with entire floors devoted to specific product lines or game platforms. Although video games dominate, many of the stores also carry a wide range of consumer electronics, action figures, T-shirts, and other memorabilia. For example, I saw several items based on the characters from Tim Burton's *The Nightmare before Christmas* (1993), in particular Jack Skellington. I had no idea that this film, which was about 10 years old at the time, was so popular in Japan. When I asked one of my guides about its popularity, she replied with one word: "Disney."

This answer surprised me, because I had not thought of *The Nightmare before Christmas* as a Disney film. Although the film appealed to children, it had several dark, quirky aspects. For example, in one scene some characters sing a song about killing Santa Claus, which is hardly Disney material. Indeed, the film was originally marketed in the United States under Disney's Touchstone label, a label used for its more mature films. Not until years later, after the film was established as a modern classic, was a 3-D version of the film released under the Disney brand name.

My exchange with the student was significant because it spoke to the success of the Disney brand in the international market generally and in the Japanese market specifically. Disney's success as a global media brand is apparent in Japan, and for good reason: Disney has had a presence there for a long time. Disney characters first appeared in Japan in 1959 on playing cards produced by Nintendo, a company that went on to become a dominant force in the video game industry.[1]

Yet, it is the success of Tokyo Disneyland (TDL) that is undoubtedly most responsible for raising the visibility of Disney in Japan. Tokyo Disneyland celebrated its 25th anniversary in 2008, and in those 25 years the park and the Disney brand have flourished. Disney is a ubiquitous presence in Tokyo, and Disney characters could be found on items for sale in several of the places I visited in the city. Even Jack Skellington had shaken off the Touchstone label and been embraced as a Disney character. This success is due to the careful integration of the Disney brand within the context of Japanese popular culture.

TDL is neither owned nor operated by Disney. The park is the property of the Oriental Land Company, which pays a licensing fee to Disney, specifically a 10 percent cut of the gate admissions and a 5 percent cut of retail sales in the park.[2] Although the park was designed to reproduce the original Disneyland in California, John Van Maanen has observed that the cultural meaning found in the park cannot be classified as exclusively American.[3] Indeed, he argues, "While Disneyland is reproduced in considerable detail, it is never deferred to entirely. . . . Japan has taken in Disneyland only, it seems to take it over."[4]

Many Disney characters have been introduced over the years, but none carry the iconic power of Mickey Mouse, who has come to symbolize the company. Van Maanen maintains that, while Mickey Mouse has primarily been associated with youth culture in the United States, the character has much broader appeal in Japan. Japanese adults have an affinity for Mickey, who is even used to market products like financial services. Van Maanen concludes that TDL reasserts Japanese culture by refiguring the meaning of American cultural icons such as Mickey Mouse, thus illustrating the complexity of global cultural flows. Indeed, TDL is an exemplar of *cultural hybridity*.

Cultural hybridity is a concept that emerged out of critiques of commercial media globalization. Some scholars of media globalization have adopted a thesis of "cultural imperialism," arguing that global media organizations perpetuate a Western value system through their media products.[5] These scholars maintain that global media organizations, driven by capitalist motives, are displacing cultural production in developing countries and in non-Western cultures, and that the distribution of Western media content is overriding the values of these cultures. In other words, the cultural imperialism thesis imagines a unilateral cultural flow that facilitates Western global dominance. This unilateral flow, it argues, has homogenized global culture in the mold of the West, and more specifically of North America.

Other scholars offer a view of media globalization that is not so unilateral. They argue that when media consumption is viewed on the local level, it is clear that audiences in developing countries do not always adopt Western cultural products nor the values they contain. Instead, elements of Western culture are sometimes reworked in non-Western localities: aspects of local culture are incorporated and new cultural products are created, which are referred to as *cultural hybrids*. For example, Stuart Hall argues that cultural hybridity is useful in understanding the diaspora experience.[6] When non-Western individuals and groups are displaced and moved into Western cultural contexts, they weave their own cultural forms into the Western forms. In these cases, cultural hybrids can appropriate Western culture as a form of resistance. Therefore, globalization can actually result in heterogeneous cultural production.

While the cultural imperialism approach to globalization sees cultural homogenization as the product of capitalist interests, the heterogeneity of cultural hybridity is often held up as an example of resistance to these interests. There is, however, another type of cultural hybridity, which emerges from the practice of *glocalization*. Glocalization occurs when international corporations tailor their products to local markets.[7] In cases of glocalization, cultural hybrids are not products of resistance to the interests of capitalism, but the products of capitalism itself.

Van Maanen's analysis of TDL illustrates the glocalization of the Disney brand in terms of the experiences and products offered in the park. Avaid Raz has found that the management of the park also reflects this type of cultural hybridity.[8] In an analysis of training programs utilized at TDL, Raz found that the "Disney Way" program (a management program used by the Disney corporation) was reworked to accommodate the Japanese workforce. The Disney organizational culture, which presents the corporation as a family of employees, is similar to the established Japanese organizational culture, so some elements were rather easily incorporated directly into TDL. For example, employees at TDL are subject to the same strict dress codes that are enforced at the U.S. Disney parks, but similar codes were already present in "the pre-existing local socialization to appearance management in Japan."[9] The managers at the Oriental Land Company, however, omitted parts of Disney's training program that they found inappropriate or unnecessary, particularly where part-time workers were concerned—workers in whom the company had little long-term interest. Managers decided that, unlike their counterparts in the other Disney parks, part-time workers at TDL did not need training in the history of Disney, nor did the training of these workers need to take place in facilities as nicely decorated as those at Disney's U.S. parks. Raz concludes, "TDL, as a case study in the hybridization of organizational culture, can also provide a window to the broader processes of globalization or rather glocalization (the local articulation of global culture)."[10]

The cultural products of Disney, as well as its management style, are refigured in TDL, but these are the products of one large corporation that are being locally articulated by another large corporation. Indeed, as Raz notes, the Oriental Land Company is actually a partnership of a real estate development company and a railway company. In the case of TDL, cultural hybridity is not an act of resistance performed by displaced or marginalized peoples; it is the product of a licensing contract between two corporations, and as such is firmly entrenched in the interests of capitalism. Disney's willingness to allow TDL to adapt and modify both its cultural products and its management style for the Japanese locale was motivated by these shared interests.

Glocalization, then, explains how cultural hybridity can serve corporate interests, and it is a practice that has served Disney quite well. The success of TDL is inherent in its hybridity, which allowed

Disney to be "taken over" by the Japanese, with an outcome that was apparent throughout Akihabara and greater Tokyo. In other words, Disney has profited significantly from Japanese consumers who embrace these American characters as their own. Although these consumers have, as Van Maanen would put it, reconstituted Mickey Mouse and given him new meaning, Mickey is still a valuable and profitable American original.[11]

Disney has extended its practices of cultural hybridization in its collaboration with the video game company Square Enix, producing a series of video games known as *Kingdom Hearts*. The *Kingdom Hearts* games are role-playing games (RPGs) that incorporate characters from Square Enix's *Final Fantasy* games and from Disney's animated films. Although the games have been published worldwide, they are particularly popular in Japan and the United States. Therefore, the games themselves are bilateral cultural hybrids, and the concepts of globalization and hybridity provide critical frames through which to view the games more closely.

I will begin by presenting background information on the RPG genre, the *Final Fantasy* series, and the *Kingdom Hearts* games. I will explain the context in which these games were released, a period of time in which the Disney brand was in decline and the Disney corporation in upheaval. I then will provide an analysis of the *Kingdom Hearts* and *Kingdom Hearts II* games to show that, while they incorporate Disney characters into narratives that put them in league with *Final Fantasy* characters, the stories are all Disney's. In fact, the story told in the *Kingdom Hearts* games portrays Disney as a benevolent force that provides a home and family for a displaced boy and fights against an evil force that threatens to destroy worlds and erase their cultures.

RPGs, *Final Fantasy*, and *Kingdom Hearts*

The RPG is a genre of video games inspired by the old *Dungeons and Dragons* game, which in turn was based on the *Lord of the Rings* trilogy. Therefore, the narratives of most RPGs are set in a fantasy context, either historical or futuristic, and the adventures are often epic in scope, traversing continents, worlds, and sometimes galaxies. Mark J. P. Wolf describes RPGs as "games in which players create or take on a character represented by various statistics, which may even include a developed persona. The character's description may include specifics such as species, race, gender, and occupation, and may also include various abilities, such as strength and dexterity."[12]

In RPGs, the game player has either a preset avatar or a set of characters from which to choose an avatar. As the game progresses, the avatar gains more experience, and this allows the player to imbue it with special skills or to enhance the abilities Wolf mentions. RPGs vary depending on the specific game, and, as Andrew Burn and Diane Carr have observed, generic differences are manifest in games that have been developed for play on the PC (*Baldur's Gate*), on game consoles (*Final Fantasy*), and online (*Anarchy Online*).[13] At this point, I must confess that, of all video game genres, RPGs are my favorite, and the *Final Fantasy* games hold a special appeal for me.[14]

The *Final Fantasy* franchise was created by Hironobu Sakaguchi, a game designer who worked for what was then Square Co., Ltd. Sakaguchi was planning to leave Square Co. but wanted to develop one last game for the company, and that game was *Final Fantasy*.[15] The game, released in Japan in December 1987, proved to be a phenomenal success, and many additional games have since been released as part of the *Final Fantasy* series.

It would be difficult to overstate the popularity that these games enjoy in Japan. In Akihabara, stores devote entire sections to the *Final Fantasy* franchise, with all of the games as well as action figures of the most popular characters available for purchase. New releases in the game series are highly anticipated in Japan, and it is not uncommon for a new *Final Fantasy* game to sell millions of units in the first week of its release.[16] In addition, the music from the games, composed by Nobuo Uematsu, has also become quite popular in Japan. Some of this music has been performed by the Tokyo Symphony Orchestra, and much of the music from individual *Final Fantasy* games is available on CD.[17]

While the RPG genre has always been popular in Japan, it has not always been as popular in the United States. The *Final Fantasy* games are credited with popularizing RPGs in the United States, and *Final Fantasy VII* is considered to be one of the first RPGs to enjoy widespread popularity in the U.S. market.[18] Not all of the *Final Fantasy* games make it to the U.S. market, however, and those that do are often released months or years after their release in Japan. Moreover, these games consistently sell more units in the Japanese market than in the United States, even though the United States is the larger market.[19]

For these reasons, the *Final Fantasy* games are produced primarily for a Japanese audience, which is clear to anyone who has played them. For example, in *Final Fantasy X,* the visual aesthetic of the game environment, the buildings, and the costumes of the characters incorporate elements of Japanese design. Despite their popularity in the United States, the *Final Fantasy* games retain these Japanese elements because their primary market is Japan. In other words, the *Final Fantasy* games reflect their country of origin, and Japanese players have embraced the games as indigenous cultural products.[20]

Yet the *Final Fantasy* games are cultural products with a corporate history. As I noted earlier, the games were originally produced for Square Co., Ltd. and that company (despite name changes) has continued to produce and publish the *Final Fantasy* games, often under the SquareSoft brand label. Square Co. had great ambitions for the *Final Fantasy* franchise, including a CGI feature film based on the game series. That film, *Final Fantasy: The Spirits Within,* was released in 2001 and failed miserably; it had an estimated production budget of $137 million but garnered only $85 million in worldwide box office receipts.[21] About two years after this failure, Square Co. Ltd. merged with Enix, the company behind another important RPG series (*Dragon Quest*), to form Square Enix. Square Enix is now the dominant force in the RPG genre, owning the *Final Fantasy, Dragon Quest,* and *Kingdom Hearts* series.

The *Kingdom Hearts* and *Kingdom Hearts II* games are collaborative projects between Disney Interactive and Square Enix, although the first game was produced in collaboration with Square Co. prior to the merger. *Kingdom Hearts* has become a franchise in its own right. In addition to the two games produced for consoles, a number of games have been produced for handheld gaming devices, including *Kingdom Hearts: Chain of Memories* for Nintendo's Game Boy Advance; *Kingdom Hearts: 358/2 Days* for the Nintendo DS; and *Kingdom Hearts: Birth by Sleep* for the PlayStation Portable. *Kingdom Hearts* games have also been developed for mobile telephone services like Verizon Wireless.[22]

Of all these games, *Kingdom Hearts* (*KH*) and *Kingdom Hearts II* (*KH II*), released on the PS2, have been by far the most successful. *KH* was released in Japan in March 2002 and in the United States in September of that year; the game sold 1.24 million and 3.45 million units, respectively, and a total of 5.5 million units worldwide. *KH II* was released in Japan in December 2005 and in the United States in March 2006; it sold 1.16 million units in Japan, 2.06 million units in the United States, and 3.96 million units worldwide.[23] These figures certainly speak to the success of these games as cultural hybrids. The games gave Disney another opportunity to profit from the popularity of its characters in the Japanese market, while Square Enix was able to extend its success in the U.S. market.

About the time *KH* and *KH II* entered the market, however, the Disney brand was experiencing some problems and the company was going through a very public imbroglio. In this context, these games offered an additional advantage for Disney, which can best be understood by examining how Disney's problems emerged.

The Rise and Fall of Disney

In November 2003, Disney celebrated the 75th birthday of Mickey Mouse, a character introduced to film audiences in the 1928 animated short *Plane Crazy*. For all practical purposes, Mickey's birth date also marked the birth of the Disney brand, and, as media brands go, Disney is an exemplar. Walt Disney worked hard to cultivate an image for his company as a trusted source of family entertainment. To achieve that trust, he cultivated an image of himself as "Uncle Walt"—not a media mogul, but a member of the family—and the Disney brand became intimately linked with Walt Disney's family-friendly persona.[24] For example, prior to opening Disneyland, Disney struck a deal with ABC to air a weekly television show with him as the host. Later, the show moved to NBC, where it became perennially popular as *The Wonderful World of Disney.* Walt Disney continued to host the show until his death in 1966, beaming his persona into American households as a means of promoting his company's products.

Apart from creating a media brand, Disney is also credited with having the vision to develop his business into the model of a diversified media company. Disney understood the significant advantages of promoting and producing media products across a variety of platforms and thus generating multiple revenue streams. After Disney's death, the company continued to follow his model, with varying degrees of success. In 1984, however, the model began to show its age, and the company was in trouble. Disney's board decided to oust president and CEO Ron Miller, replacing him with the team of Michael Eisner and Frank Wells. Eisner, as chair and CEO, headed creative production at Disney, and Wells, as president and COO, handled the business end. Ten years later, Wells was killed in a helicopter accident, and Eisner took complete control of the company. Ten years after that, Eisner was forced out of Disney by a shareholder coup.

Although Eisner can be credited with much of Disney's success during the years of his leadership, he was also responsible for some of its problems.[25] For example, Disney purchased the art house imprint Miramax in 1993, but the studio continued to make production decisions independently of Disney. Some of those decisions included producing films that offended social conservatives, such as *Priest* (1994), a film about a Catholic priest struggling with his homosexuality. Because Disney was ultimately responsible for releasing Miramax films, its image as a source of family entertainment was called into question. Socially conservative political and religious organizations claimed that the company no longer embraced "family values," and boycotts against Disney were organized by the Catholic Church and the Southern Baptist Convention.[26]

Eisner was responsible for other decisions that further eroded the Disney brand. Eisner was slowly phasing out Disney's animation studios, cutting back sharply on the very type of film production that had made the company famous. In 1999, Disney employed 2,220 people in its animation division; by 2004, that number had dropped to 600 and the company had closed animation studios in Paris, Tokyo, and Orlando.[27] At the time, Disney's most successful animated films were the collaborative projects with the computer animation studio Pixar. However, Pixar's contract with Disney was up for negotiation, and those negotiations were not going well. Steve Jobs, who owned Pixar, was vocal about what he saw as the inequities in the contract Pixar had with Disney, which gave the latter a greater share of the box office receipts for Pixar's films. Through most of 2003, Jobs and Eisner had been trying to negotiate a new agreement between the two companies; Jobs wanted changes in the contract, but it appeared that Eisner would not budge.[28]

On December 1, 2003, Roy Disney, a vice chair and nephew of company founder Walt Disney, resigned from the Disney board.[29] He did not go quietly. Rather, he used the occasion to criticize Eisner publicly, issuing a three-page letter in which he held Eisner responsible for the flagging performance of the company and called for his resignation. Roy Disney's resignation also brought to light internal conflicts within the company and the building resentment of Eisner's management style. The next day, another board member, Stanley Gold, resigned, seconding Roy Disney's call for Eisner's resignation.[30]

Eisner suffered another public embarrassment when Jobs announced on January 29, 2004, that he was ending the negotiations to continue the collaboration between Pixar and Disney.[31] Given that other major studios were eagerly courting Pixar, Roy Disney wasted no time in using the event to support his call for Eisner to step down. In the months that followed, he mounted a "Save Disney" campaign, rallying Disney stockholders to remove Eisner and to restore the company's image as a purveyor of family-friendly entertainment.[32] Eisner was held up as an evil presence who threatened Disney's survival and who must be banished from the kingdom in order to save Disney.[33] The campaign worked. In March 2004, the board of directors at Disney took the chair position away from Eisner and gave it to another board member, George Mitchell. Eisner stayed on for another year as Disney's CEO, but stepped down from that position in March 2005; Bob Iger was chosen as his replacement.

When Iger took over at Disney, he successfully negotiated with Jobs to buy Pixar outright. In 2006, Disney acquired Pixar and Jobs joined the Disney board. Pixar still produces films under its own studio name, and those films continue to be successfully distributed by Disney. Under Iger, Disney has also reasserted itself in the market as a family-friendly entertainment brand, specifically targeting young girls with the Miley Cyrus/Hannah Montana, Jonas Brothers, and *High School Musical* franchises. In the years in which the *KH* and *KH II* games were developed and released, however, the Disney brand was in trouble. In the analysis that follows, I will show that *KH* and *KH II* were designed to reestablish Disney's corporate identity and reclaim its image as a benevolent cultural presence.

Hearts and Darkness

Similar to other games I have analyzed, once the game disc is inserted, *KH* requires the player to initiate an action on the start-up screen, either choosing a new game or loading an ongoing game. If the player does not make a choice, a trailer begins to play. Displaying gold text on a black screen, the trailer begins: "There are many worlds, but they share the same sky—one sky, one destiny." Cut scenes from the game then play, showing the different characters and the various worlds in the game. These scenes depict a variety of Disney characters but focus primarily on the main character, Sora. Sora is an adolescent boy who serves as the avatar for the player. Much of the game play involves Sora, who is also the protagonist in the narrative. This opening cut scene foreshadows an important point that is revealed later in the game. The Magic Kingdom emerges as the "one sky, one destiny" that these worlds share, and the Disney characters emerge as the defenders of these worlds.

Once the player has initiated a new game, a cut scene begins with Sora's voice-over: "I've been having these weird thoughts lately. Like is any of this for real or not?" The words are spoken over images of Sora floating in an abstract space that appears to be a body of water. The spoken sentences flash on the screen, and the words *real* and *weird* linger after the rest have disappeared. Another scene then plays, showing Sora emerging from the sea to greet his friend Kairi, who then reacts to an image of Sora falling from the sky. Sora plunges into the depths of the water, landing in a darkened space illuminated only by a stained-glass floor depicting Snow White and the Seven Dwarfs, the first in a series of stained-glass floors depicting Disney heroines. The player later learns that these stained-glass images depict the Princesses of Heart, who do not harbor darkness within their hearts. These are all Disney princesses, heroines from various Disney films, including Snow White and Cinderella. Although not every Disney princess appears in the game, many do; their presence is worth noting because these princesses function as a franchise for Disney, and the company markets numerous products based on these characters. Disney even has an official web page for its princesses that promotes the products, films, and live events in which they are featured.

At this point in the game, the player must choose a weapon that determines Sora's skill level at the beginning of the game. After a tutorial and a boss fight, Sora wakes up and explains to his friend Kairi that he just had a strange dream.[34] As they discuss his dream, it is revealed that Kairi is a refugee, displaced from her home at such a young age that she cannot remember much about it. She has clearly found a new home with Sora and their friend Riku. Riku interrupts this conversation to ask why no one is working on the raft they are building. The cut scene concludes with the *KH* logo appearing on the screen, followed by the logo for this particular world, Destiny Island.

This island is Sora's home, which is populated by his adolescent friends Riku and Kairi and the *Final Fantasy* (*FinF*) characters Tidus and Wakka (from *FinF X*) and Selphie (from *FinF VII*). But these *FinF* characters are rendered as younger, adolescent versions of the characters that appeared in the original games. Those familiar with the *FinF* games will recognize that Sora and his friends share some of the features of many of the *FinF* characters. For example, both Sora and Riku have the spiky hairstyles and complex outfits typical of *FinF* heroes. It is unclear exactly what Tidus, Wakka, and Selphie are doing on Destiny Island, but their presence serves to identify Sora with the *FinF* franchise. Kairi, in comparison to Sora and Riku, is dressed and styled in a less garish way, and for good reason. Later in the game, it is disclosed that she is one of the Princesses of Heart. Thus, while Sora's and Riku's appearance associates them with characters in the *FinF* games, Kairi is associated with Disney characters.

Destiny Island appears as a tropical isle in the middle of a bright blue sea, with tree houses, palm trees, and exotic fruit. Although it would seem to be an adolescent's paradise, Sora and his friends are building a raft so that they can leave the island to explore and perhaps find Kairi's home. In a conversation among the three friends, Riku speculates on the randomness of their presence on the island, musing, "I've always wondered why we're here on this island. If there are any other worlds out there, why did we end up on this one? And suppose there are other worlds, then ours is just a little piece of something much greater. So we could have just as easily ended up somewhere else, right?" Sora replies, "I don't know," to which Riku responds, "Exactly, that's why we need to go out there and find out."

In this exchange, it seems that Riku and Sora are ambivalent about their home, and their presence on Destiny Island seems as random to them as Kairi's presence. Although the name "Destiny" suggests the point at which a journey ends, it is actually a point of departure. Sora calls it home, but within the narrative of the game he spends little time there. And once he leaves—which is soon—Sora's only attachment to the island is his relationship to Kairi and Riku.

The scene then shifts to the Magic Kingdom, where Donald Duck finds a note left for him by King Mickey. In this note, Mickey says he has observed that "stars have been blinkin' out, one by one. And that means disaster can't be far behind. I hate to leave you all but I've gotta check into it." King Mickey also instructs Donald and Goofy to find "someone with a 'key'—the key to our survival," and he directs them to Traverse Town to seek out Leon for help. The "key" to which Mickey refers is the Keyblade, a weapon that will soon fall into Sora's hands. Leon is a character who appeared in the *FinF VII* game as the hero Squall Leonhart, and Mickey's knowledge of him foreshadows the relationship that develops between the Disney characters and the *FinF* characters in the *KH* games.

The scene now shifts back to Destiny Island, where Sora is awakened by what appears to be a storm. The storm is actually a dark force that threatens to consume the island, and it brings with it the Heartless. The Heartless are the minions of this dark force, creatures that were once human but that have lost their hearts to the darkness. Now, they are faceless creatures with no distinctive features or discernible identities. Although Sora will encounter various kinds of Heartless creatures in the worlds he visits, all of these creatures (with the exception of specific "bosses" he must battle) are faceless and lack characteristics that distinguish them from one another.

As Sora unsuccessfully battles the Heartless, he finds Riku, who tells Sora that "a door has been opened to the outside world." Riku states that, although he may never be able to return to Destiny Island or see his family again, it is a chance to escape and he "is not afraid of the darkness." Riku then reaches out for Sora but is engulfed by the darkness. Sora finds himself alone to face the Heartless, but at this point a mysterious voice speaks to him and the Keyblade appears in his hand. The Keyblade is a weapon in the shape of a skeleton key, but Sora wields it like a sword. He uses the weapon to fight the Heartless and a boss (a larger manifestation of the darkness), but ultimately he is unable to save his home, which is destroyed by the darkness.

This dark force is what King Mickey observed earlier, and it is revealed later that it is this same force that destroyed Kairi's home and left her a refugee on Destiny Island. It is important to note how this force is portrayed in the game: it destroys worlds, erases their cultures, and makes their inhabitants homogenized, faceless, Heartless creatures. In other words, this dark force manifests the negative effects of cultural imperialism.

After Destiny Island is taken by the darkness, Sora wakes up in what he learns is Traverse Town. As the name implies, Traverse

Town is a place people pass through, and in the game it serves as the departure point for Sora's trips to other worlds. A barmaid highlights this fact when she tells him that everyone in Traverse Town is from someplace else, and she implies that bad events have brought people to the town. The same can be said of Sora, and at this point it is clear that Traverse Town is a diaspora, and Sora is now also a refugee. It seems that most of the populace of Traverse Town have had their worlds destroyed as well, and some of those present seem to be refugees from the worlds of *FinF* games.

In Traverse Town, Sora meets Leon and other *FinF* characters, including Yuffie and Aerith from *FinF VII,* whose worlds have all been destroyed by the Heartless. *FinF* characters also appear elsewhere in the game; for example, Sephiroth, the infamous villain from *FinF VII,* is the final boss in the "Coliseum Tournaments" challenges. Traverse Town, however, is a world in which the Disney and *FinF* characters live and work together. Cid, a character who has appeared in several *FinF* games, runs a shop in the town, as do Donald Duck's nephews Huey, Dewey, and Louie. The player uses these shops to stock up on needed items and to upgrade equipment. Although these characters come from different worlds, they all—even the *FinF* characters—acknowledge Mickey as the king and refer to him as such. The Disney and *FinF* characters now occupy the same world and share the same story, as these cultural brands converge under the crown of the Magic Kingdom.

When Leon and Yuffie meet Sora, they explain to him what they know about the Heartless, and they also explain that the Keyblade has chosen him to be its master. At this point, Sora realizes that his own home may have been destroyed by this force. Yuffie asks Sora if he has heard of Ansem, explaining that Ansem disappeared while studying the Heartless. Ansem recorded his findings in a report before his disappearance, but that report has been scattered across different worlds.

Meanwhile, Aerith meets Donald and Goofy, and during this exchange Aerith explains how the various worlds are being connected and destroyed by the darkness. Donald and Goofy express surprise, primarily because the knowledge of the different worlds was supposed to be a secret. In other words, while the Magic Kingdom was aware of these different worlds, the separate worlds had no knowledge of each other until now, when they are being connected by the darkness and ultimately destroyed. Again, this knowledge of the different worlds indicates that King Mickey is at least all-knowing, if not exactly all-powerful.

Sora also meets Donald and Goofy in Traverse Town, and together they battle the Heartless. After winning a boss battle with another Heartless, Sora decides to team up with Donald and Goofy to search for King Mickey, Riku, and Kairi. But before Sora can join them, Donald insists that he put on a "happy face"—as if they were recruiting him to be a Mouseketeer—and they pledge "one for all, and all for one." At this point, Sora has found new friends in the diaspora of Traverse Town. Given that his previous home was defined more by his friends than his family (who never actually appear in the game), Donald and Goofy's friendship offers Sora the possibility of a new sense of home. Their only demand of him is that he must at least display the outward appearance of happiness, a hallmark of the Disney brand.

Although Sora has found new friends, it appears that he also has made new enemies. The scene shifts to a meeting of famous villains from Disney films, including Hades from *Hercules* (1997) and Ursula from *The Little Mermaid* (1989). Led by Maleficent (the sorceress from *Sleeping Beauty,* 1959), they discuss how to handle the threat that Sora poses.

Indeed, they will all encounter Sora as the game progresses, as he, Donald, and Goofy visit the various worlds and eliminate the Heartless. All these worlds represent Disney films, and they include Atlantica (from *The Little Mermaid*), Neverland (from *Peter Pan,* 1953), and Deep Jungle (from *Tarzan,* 1999). In each of these worlds, Sora, Donald, and Goofy must learn the terrain, battle the Heartless, and help the Disney protagonists to battle the various villains. In this way, Sora interacts with the Disney characters and participates in the narrative conflicts from the films these worlds represent.

For example, one of the worlds Sora visits is the Olympus Coliseum, and there he meets Hercules and his trainer Philoctetes, or Phil. In this world, Sora helps Hercules to battle his nemesis Hades, who is in league with the Heartless. The character Cloud also appears in this world. Cloud originally appeared as the main character in *FinF VII,* but here he is a mercenary for Hades. Sora must battle Cloud before he can assist Hercules in the battle against the three-headed Cerberus, a monster created by Hades. In another world, Agrabah, Sora meets Aladdin and helps him to free Jasmine from the clutches of the villain Jafar. In Agrabah, the game play is woven into the narrative offered in the film *Aladdin* (1992), and the player can even choose to have Aladdin fight alongside Sora in the battle against Jafar.

At the conclusion of the final battle in each world, Sora uses the Keyblade to close the portal that has been used by the Heartless to enter these worlds, and by doing so he closes the connections these worlds had with the dark forces that devoured his own. Thus, Sora, Donald, and Goofy are on a mission, entrusted to them by King Mickey, to battle the destructive, homogenizing effects of the Heartless.

Sora remains transient throughout the game, a character with no home, though he forms important relationships within the worlds he visits and saves. In fact, his transience even changes his identity to the degree that his appearance morphs to adapt to some of the worlds he visits (as do those of Donald and Goofy). For example, when he goes to Atlantica (from *The Little Mermaid*), Sora becomes a merman with a fin where his legs once were; when he goes to Halloween Town (a world that draws on *The Nightmare before Christmas*), he adopts a "scary" costume much like the other residents of this world. In some of these worlds, the transformation appears to serve a practical function; in Atlantica, for instance, his fin allows him to swim in the sea. In other worlds there appears to be no practical reason for the transformation other than aesthetics. In each of these instances, however, Sora's physical appearance—his identity—changes to adapt to the worlds he ultimately saves.

One of the final worlds Sora visits is Hollow Bastion, a place destroyed by the Heartless that turns out to be Kairi's home. Sora confronts and defeats Maleficent and seals this world away from the Heartless, but the game is not over. Sora's next battle occurs at "The End of the World," where the final scenes and battles of the game play out. Here, Sora encounters Ansem and learns about Kingdom Hearts, a place where all hearts are thought to be born.[35] Kingdom Hearts is represented as a door, and Ansem explains that darkness resides behind the door, but within the darkness is the force of light.

If Kingdom Hearts seems like an obvious metaphor for the idea that good and evil reside in everyone, that is because it is. Indeed, this idea is present in other aspects of the game as well. The character Riku, for example, willfully gives himself up to the darkness that attacks Destiny Island, then reappears later to join forces with Maleficent. His jealousy of Sora and Sora's status as the keeper of the Keyblade reaches a point that he turns on his friend and tries to kill him. Yet, in an act of self-sacrifice in the final stages of the game, Riku actually helps Sora in his fight against the Heartless.

Sora, too, displays the influence of both good and evil when he briefly becomes a Heartless toward the end of the game. Yet some characters in the game are clearly consumed with evil. Ansem, for example, has become so fully a creature of darkness that he is destroyed by the light that emerges from the opened door to Kingdom Hearts. King Mickey, in contrast, clearly represents good, and he appears in the final stage of the game to help Sora defeat the Heartless. And, at the end of the game, it is Mickey, with the help of Riku, who seals himself behind the door to Kingdom Hearts so that the other worlds will be saved from darkness. Then, of course, there are the Princesses of Heart mentioned earlier—Disney princesses with hearts filled with goodness and light. In fact, it seems that whatever power rests within Kingdom Hearts is fundamentally good, though characters like Ansem seek to exploit its darker elements.

KH II begins with a new character, Roxas, serving as the avatar, replacing the protagonist Sora.[36] As the game progresses, it is revealed that Roxas is the Nobody of Sora, and Nobodies are another new element. Nobodies are the shells of people who have fallen victim to the darkness and become Heartless, as Sora did for a moment during the first game. Early in *KH II,* Sora reunites with the Nobody Roxas and restores his full identity, and the rest of the game is played with the character Sora as the avatar. The Nobodies, however, also comprise Organization XIII, a group that appears in this game as the main antagonists who battle Sora and his allies over the power of Kingdom Hearts.

I should mention at this point that the *KH* franchise's narrative sometimes teeters under the weight of its own metaphysics; performing the necessary exegesis to explain how all these elements fit together in the internal logic of the narrative would take me far afield. It should suffice to say that certain important elements from the first game carry over to *KH II,* and darkness still threatens the worlds that represent the Disney films. In addition, Sora, Donald, and Goofy are still defending these worlds against this dark force, and they are also still on a journey to find King Mickey and bring him safely back to the Magic Kingdom.

KH II introduces several new worlds not included in the first game. The choice of some of these worlds seems obvious. For example, *KH*

II includes the Pride Lands world, based on Disney's *The Lion King* (1994). *The Lion King* is one of Disney's most successful films and once held the record for the highest-grossing animated feature. In the Pride Lands, Sora helps Simba to fight Scar, so the action of the game closely follows the narrative of the film. In addition, Sora, Donald, and Goofy all change appearance when they enter this world, adopting the physical characteristics of the animals that appear in the film.

Port Royal is another new world introduced in *KH II,* drawn from Disney's live-action feature *Pirates of the Caribbean: The Curse of the Black Pearl* (2003). In this world, Sora meets up with Jack Sparrow, Will Turner, and Elizabeth Swann; he also battles Captain Barbossa and his undead minions. Again, the choice of this world seems obvious given the success of the *Pirates of the Caribbean* franchise; the three films have grossed over a billion dollars. It appears that with *KH II,* Disney decided to exploit some of its most successful properties, one of which (*Pirates*) was not yet released when the first *KH* game hit the market.

KH II also includes worlds that reflect less popular and less lucrative Disney properties. For example, the game includes the Timeless River world, which attempts to recreate the experience of Disney's earliest animated films. This world is rendered in black and white, and the characters, including Sora, morph to reflect the wide-eyed, thin-limbed style in which many early animated characters were rendered. More specifically, Timeless River references some of the films that first introduced Mickey Mouse, including *Steamboat Willie* (1928) and *Plane Crazy.* In this world, Sora must battle Pete, a character who appeared in many early Disney films as Mickey's perennial antagonist.

Space Paranoids is another new world in *KH II,* based on the Disney film *Tron* (1982). *Tron* is notable not for its commercial success but for its use of computer-generated animation, which at the time was considered state-of-the-art. The film tells the story of a video game designer whose work is stolen and who becomes digitized and trapped in a mainframe computer by the man who stole his designs. *Tron* was one of the first films to not only draw on the aesthetics of video games, but also incorporate them into the narrative. In *KH II,* however, the computer in question is now Ansem's, and Sora, again physically transformed by the world, must battle a "hostile program" to stem the tide of the dark forces.

Obviously, the inclusion of these new worlds extends the *KH* story merely by adding in more of Disney's intellectual property, and these new worlds reveal an irony in the grand narrative of the *KH* games. Although Sora, Donald, and Goofy are on a mission to defend these worlds from destruction and homogenization, the worlds are all the intellectual property of a single corporation: they do indeed all "share the same sky."

As *KH II* progresses, Sora learns that the Ansem he battled in the first game was not, in fact, Ansem. Instead, it was the Heartless character Xehanort, a former student of Ansem (who is referred to in *KH II* as "Ansem the Wise"). In *KH II,* Ansem is portrayed as a benevolent scholar whose research on the Heartless was misused for evil by his rogue student, Xehanort. It is Xehanort's Nobody, Xemnas, whom Sora must defeat, because Xemnas is the leader of Organization XIII (again, the metaphysics of *KH II* are, at times, staggering). Sora travels to the Organization XIII headquarters, "The World That Never Was," and confronts and defeats Xemnas with help from King Mickey and Ansem the Wise. After winning this battle, his home of Destiny Island is restored, and there he meets up with his friends Kairi and Riku. During one of the final cut scenes, the three receive a letter from King Mickey indicating that another adventure, and another sequel to the game, lies ahead.[37]

■ ■ ■ In the *KH* games, Sora's story is that of a refugee. His home is destroyed by a dark, alien power, and he is displaced and separated from his friends Kairi and Riku, who seem to have more presence in his life than does his biological family. He ends up in Traverse Town, a place populated by refugees from other destroyed worlds. While in this place, he meets two new friends, Donald and Goofy, and the three set off on a quest to save the other worlds, which are threatened by the darkness that wishes to connect and ultimately consume them. When they arrive in some of these worlds, they adapt themselves to the environment by changing their physical appearance. After successfully defeating the legion of darkness and protecting these worlds, Sora returns to a restored Destiny Island to reunite with Kairi and Riku.

Sora's story is not unique, as many RPGs chronicle the struggles of displaced individuals searching for their homes (Tidus in *FinF X,* for example). Yet the resolution of the conflicts in Sora's story reveals that it is really a story about Disney. Wherever he travels and whatever identity he adopts, Sora is still within the worlds of Disney. In fact, the moral of this particular story suggests that those who have lost their homes can find one in Disney. Moreover, those who trust in King Mickey have help in their fight against evil and may even win back the home they have lost.

The story of *KH* functions as a parable about globalization. Worlds are connected by darkness and their citizens are turned into faceless, Heartless creatures. It is difficult to ignore the parallel with the critique of cultural imperialism, which imagines Western culture as a dominant force that erases cultural distinction. It is also important to note that the destruction of these worlds occurs *once they are connected,* in much the same manner as the transglobal communication networks established by the major media conglomerates have connected once-separate worlds.

In light of these parallels, Disney occupies a curious role within the games as the defender of worlds against the threat posed by this dark network. It is King Mickey who first recognizes the threat. The Magic Kingdom clearly had foreknowledge of the worlds, and King Mickey uses this knowledge to monitor their safety; when danger threatens, he springs into action. It is he who dispatches Donald and Goofy to seek out the keeper of the Keyblade, and they, as agents of King Mickey, who induct Sora into their quest to save the worlds.

On this point, the irony is thick. Indeed, when it comes to globalization, Disney has always been ahead of the curve. Disney's entry into the Japanese market predated the corporate consolidations of the 1980s that led to the formation of other global media conglomerates. As a result, Disney served as the blueprint for the synergistic practices on which these subsequent media conglomerates were based.[38] Disney has understood for years that it *is* a small world after all or, in the case of the *KH* games, a small universe. According to the narrative of *KH,* however, the Magic Kingdom never abused its knowledge or connections and remained committed to battling against those who would. The irony gets thicker when we consider that most of the worlds that Disney protects in this game are based on its own intellectual properties. In other words, in *KH,* Disney actually owns the worlds it is defending.

It is also important to note how, in the context of the game, the knowledge of these worlds is misused by those corrupted by power. Xemnas, after all, gained his knowledge from Ansem the Wise, and this relationship has a parallel to Disney's corporate context. Eisner was never a student of Walt Disney and Walt never served as his mentor, but there was an effort at Disney to represent Eisner as the new Uncle Walt. For example, in 1988 *The Wonderful World of Disney* was resurrected as *The Magical World of Disney*, with Eisner as the new host.[39] In other words, Eisner attempted to assume Uncle Walt's role in a manner similar to the way Xemnas assumed Ansem's identity. Like Xemnas, Eisner was characterized in the "Save Disney" campaign as corrupting the power of the Disney corporation, and also like Xemnas, Eisner was portrayed as corrupted by power. Certainly, Eisner would have been unlikely to allow the release of these games if he imagined they portrayed him in a negative light. At the very least, however, the story told in the games is fundamentally the same story spun about the battle over the Disney company: a tale of good triumphing over the corrupting force of power.

The *KH* games are cultural hybrids, and they successfully appeal to players in both the U.S. and Japanese markets. In some ways, the games are an extension of the same practices of glocalization found in TDL. Yet Square Enix also benefits from these practices by incorporating its characters into a game with the Disney brand, which is well recognized in the United States. Given the success of the games in both the U.S. and Japanese markets, it seems that both Square Enix and Disney have enjoyed the profits of this particular hybrid.

Yet Disney enjoyed another advantage because it used the games to tell a story about itself, a story in which the dark force that once threatened the worlds of the Magic Kingdom was ultimately contained and everything was returned to normal, including the adolescent paradise of Destiny Island. No matter who might attempt to exploit Disney for evil purposes, the story tells us, in the end the inherent goodness and light at the heart of Disney will prevail. I am sure Uncle Walt would not have had it any other way.

SIX

WHAT SHALL WE PLAY NEXT?

Throughout this book, I have analyzed games that were produced for the PS2 console. As I noted in the first chapter, the PS2 is one of the most successful gaming consoles in the history of the video game industry. It appeared among a generation of consoles that included Microsoft's Xbox and Nintendo's GameCube, and it outsold both by millions of units. Nevertheless, the days of the PS2 are waning, and its replacement, the PS3, has already been available for years. A new generation of consoles now dominates the market, including the PS3, Microsoft's Xbox 360, and Nintendo's Wii.

The history of the video game industry can be articulated by the progression of game console technology, and this progression can be demarcated by the different generations of consoles. For example, older generations of consoles were referred to by their processors' capacities (8-bit and 16-bit systems) and were marketed on their ability to display ever more detailed graphics. Even today, new consoles are marketed in this way. The video game industry furthers itself through this technological advancement, which has been woven into the industry's marketing schema: new is better. Therefore, new video game consoles are always pitched as offering the most advanced technology and thereby delivering a better gaming experience. This is an industry that focuses as much on the future as on the present. Video game blogs are always speculating about what lies just over the horizon in terms of new games and gaming systems. The video game press and its audience seem to exist in a state of perpetual anticipation of what will be played next.

In the previous chapters, I analyzed specific games to show how their designs have addressed certain corporate conditions. In this chapter, I will explore some of the technologies that relate to film and video game convergence and some of the issues that could be addressed in future studies of this convergence. I will begin with an analysis of the competition between the PS3 and the Wii. These consoles offer different types of new technology, and different strategies have been used to introduce these technologies into the video game market.

I will then discuss how advancements in fiber optic telecommunications technology have led to an increase in online gaming, specifically in massively multiplayer online role-playing games (MMORPGs). I will discuss the unique cultures of MMORPGs and how these cultures pose problems for studios that want to repurpose film content for online games. Finally, I will explore how the film and video game industries are working closely with the military to develop new technologies. This specific collaboration has broad political implications that could inform future studies of convergence. All of the issues discussed in this chapter have implications for the practices of convergence and for the future of these practices.

Console Wars/Format Wars

I visited Akihabara during the Christmas season of 2005, only a few days after the Xbox 360 was released in the Japanese market. Although in the largely Buddhist and Shintoist country of Japan Christmas is not regarded with the same enthusiasm that it is in the United States, Akihabara was still full of shoppers, most of whom seemed indifferent to the Xbox 360. It was an interesting contrast, given that, at the same time in the United States, the Xbox 360 was the coveted gift of the 2005 Christmas season and was therefore in short supply. In Akihabara, stacks of the console were everywhere, but the display units were generally ignored by the passing shoppers. This indifference is manifested in the sales figures: as of December 2009, the Xbox 360 had sold only 1.19 million units in the Japanese market, which compares unfavorably to the 3 million PS3 units sold and the 8 million Wii units sold.[1] The performance of the Xbox 360 is especially poor given that it was released a year prior to both the PS3 and the Wii. Clearly, the Microsoft brand does not enjoy the same loyal following in the Japanese market as Sony and Nintendo do.

Outside of Japan, the Xbox 360 has performed quite well, and it has sold more units worldwide than the PS3. The Wii and the PS3, however, make for a more equitable comparison because they were released at approximately the same time and both are products of Japanese companies. Yet another reason to compare these consoles is that such a comparison may illustrate the limitations of film and video game convergence. Sony has a long history as a multimedia conglomerate, and the PS3 is a product developed to exploit the synergies between Sony's film and video game companies. In contrast, Nintendo has a long history in the video game industry specifically, and its Wii was developed and marketed with a very different business strategy. In the sections that follow, I will provide brief histories of both companies and analyze the different business strategies that were used by Sony and Nintendo to develop and market their respective consoles

Sony's Synergy

While Sony seems to be a model of strategic convergence, particularly in regard to film and video games, the history of the company provides a different perspective. Sony entered into both industries reluctantly, and both entries subjected Sony to some very public humiliation. Indeed, it was the failure of Sony's Betamax that motivated Sony to acquire a film studio. Betamax was Sony's tape format for the home video market, and it lost that market to the VHS standard. Film studios supported VHS and shunned Betamax, and soon Betamax tapes were relegated to isolated shelves in the video rental stores. After this failure, Sony's leadership believed that to sell hardware the company needed to provide software, and to provide software it needed content. It was this belief that motivated Sony to hunt for a film studio.[2]

In 1989, Sony closed a deal to buy out Columbia Pictures Entertainment, which included Columbia Pictures studios, Tristar films, and the Loews theater chain.[3] This deal would allow Sony to acquire both film content and the rights to several television shows produced by Columbia. Although the acquisition would satisfy Sony's hunger for content, the deal was expensive and complex. Although Columbia was burdened with $1.6 billion in debt, Sony paid well over the market value for its shares.

In addition, Sony wanted to hire Peter Guber and John Peters to run the studio; however, Guber and Peters's production company was already under contract with Warner Studios. After Sony bought the Guber-Peters Company for $200 million, it had to spend another $800 million to settle a lawsuit filed by Warner.[4] In the twenty-first century, however, Sony's investment has paid off. Sony Pictures had the highest gross receipts of any studio in 2006 and was ranked fourth in 2007 and third in 2008. Thanks to its collaboration with Marvel Studios on the *Spider-Man* franchise, Sony can claim 3 of the top 25 highest-grossing films of all time.[5]

In terms of the video game industry, Sony's entry, if not itself a cause for embarrassment, was certainly motivated by an embarrassing event. Sony had worked with Philips to develop CD technology for the music market, and this technology went on to become the recording industry standard. Nintendo was interested in developing a new gaming system that would use a cost-efficient CD-ROM drive instead of the ROM cartridges that were in use at the time. Nintendo began collaborating with Sony to develop this new system; however, Nintendo's management became concerned that the contract outlining the collaboration gave Sony too much control. Unbeknownst to Sony, Nintendo negotiated a more favorable contract with Philips to develop the CD technology instead. At the 1991 Consumer Electronics Show, Nintendo announced its new partnership with Philips, completely blindsiding Sony.[6]

Despite this setback, Sony went ahead with the gaming project, due to the efforts of Ken Kutaragi, the project's engineer, who convinced Sony's management to move forward. As Richard Gershon and Tatuomu Kanayama note, Sony's management was reluctant to get involved in the video game industry, but finally gave Kutaragi $50 million in start-up money.[7] Sony continued to develop the technology, and out of this effort came the first Sony PlayStation gaming console, released in Japan in 1994 and to all other markets in 1995.

The PlayStation was very successful, and over 100 million units were shipped worldwide.[8] This console, however, was not the convergence technology that Sony's next console would prove to be. In 2000, Sony released the PlayStation 2 (PS2), a gaming system that offered faster processor speeds, significantly improved graphic capabilities, and a DVD player. The PS2 allowed Sony to take its film content and repurpose it for video game releases, as it did with the *Spider-Man* films, the *Men in Black* films, and even the *Stuart Little* films. In this manner, the PS2 allowed Sony to exploit licensing fees for its films in ways that other studios could not.

These same synergistic practices informed the development and marketing of Sony's later products, specifically the PlayStation 3 (PS3) and the Blu-ray DVD format. The PS3 was designed with a powerful processor capable of delivering hi-def graphics.[9] Consequently, this new system was equipped with a Blu-ray disc drive, technology that Sony marketed as the next generation of hi-def DVD. Unfortunately, Sony faced stiff competition in the video game market because the PS3 price ($499) was significantly higher than those of either the Xbox 360 ($349) or the Wii ($249).[10] Some industry analysts argued that the presence of the Blu-ray drive increased the price of the PS3, and although Sony dropped the retail price of the console, this reduction resulted in significant quarterly losses in the company's game division.[11]

Sony also faced competition in the home video market from Toshiba's entry into the hi-def format, HD DVD. Again, price was a factor, because Toshiba brought its HD DVD players to market at a significantly lower price ($229) than Blu-ray players ($439).[12] In terms of software, however, Sony had the advantage, with its Blu-ray discs usually retailing at lower prices than Toshiba's discs.[13] Toshiba sold more stand-alone disc players, but the figures for stand-alone players did not take into account the fact that the PS3, in spite of the competition from other consoles, had sold over 5 million units worldwide by the end of 2007. Toshiba's representatives argued that the PS3 sales were immaterial because consumers were not using the PS3 to play Blu-ray discs, and they estimated that only 20 percent of PS3 owners used the console to play movies.[14] This was clearly a case of Toshiba's public relations department being too clever for its own good, because by the logic of its own argument, Toshiba was claiming that 1 million PS3s were being used to play Blu-ray movies. In any event, the PS3 gave Sony an advantage in the early stages of the format war.

Another of Sony's advantages was its ability to deliver content for the Blu-ray format and its ability to ensure that some of the content remained exclusive to the format. Learning a lesson from the Betamax failure, Sony lined up content, and in addition to Columbia,

Sony acquired MGM in anticipation of the Blu-ray launch. Although MGM had slipped considerably from its days as the blue-chip studio of the old Hollywood studio system, Sony nevertheless acquired some important properties through this purchase, including the James Bond franchise. Indeed, when *Casino Royale* (2006) was released for the home video market, it appeared on DVD and Blu-ray, but not the HD DVD format. In addition to Sony's studio holdings, Disney and 20th Century Fox decided to release on Blu-ray exclusively. Toshiba had its own camp of studios, with Universal and Paramount supporting HD DVD exclusively. However, because Sony owned its studios, it could ensure exclusivity from Columbia and MGM, while Toshiba merely had agreements with Universal and Paramount.

Indeed, at the beginning of 2008 those agreements began to unravel. Although it appeared by some accounts that Toshiba was moving hardware, software sales were a different story. By October 2007, Blu-ray discs were outselling HD DVD by a two-to-one margin.[15] Where the studios were concerned, software sales mattered most, because studios were only interested in how their films performed in the home video market in these new formats. Warner, which had been releasing its films in both formats, announced early in January 2008 that it was dropping the HD DVD format. Soon after this announcement, both Netflix and Walmart decided they would no longer carry HD DVD discs. In light of these announcements, it was clear that the format war was turning in Sony's favor, yet it was surprising how quickly it did turn. In February 2008, about a month after Warner's announcement, Toshiba decided to discontinue the HD DVD format, and with that announcement Sony's Blu-ray became the industry standard for hi-def content.[16]

Clearly, Sony's advantage in the format war was its ability to strategically converge its film and video game interests. It used its studios to issue popular content exclusively to the Blu-ray format, and it used its established presence in the video game sector to get Blu-ray drives into households by loading them into the PS3. In spite of Toshiba's claim that few of these gaming systems are used to play Blu-ray movies, a report by the Entertainment Merchants Association showed that 87 percent of PS3 owners used their consoles in this way.[17]

In contrast, while Toshiba had Microsoft's support for the HD DVD format, Microsoft decided not to install HD DVD players in its

Xbox 360. Instead, Toshiba manufactured an add-on HD DVD player for the Xbox 360, but few owners of the console purchased the add-on.[18] Toshiba's failure to position its product in the video game market hurt the company in the home video market. Sony, on the other hand, built itself into the type of media conglomerate that can market new technologies competitively. Sony's decisions to acquire film studios and content and to pursue the video games sector were made with an eye to future competition. Indeed, had Sony decided not to compete with Toshiba, it would have undermined the synergistic strategy that justified the acquisition of both Columbia and MGM.

Sony's synergistic strategy becomes even more apparent in its licensing practices. Both Toshiba and Sony set up licensing structures for their respective formats.[19] The licensing fees gave producers who developed and manufactured Blu-ray or HD DVD products the right to use the format and display the respective logos on those products. Different fees were assessed for different products, depending on whether the license was for hardware or software and for recordable or read-only media. In addition, game consoles carry licensing fees, and Sony's PS3 is no exception: any game that is produced for the console must pay Sony for the rights to the system programming and the system logo. Finally, any film content from any of Sony's studios also requires a licensing fee when it is developed into a video game.

On this point, the release of *Spider-Man 3* is instructive. Although Sony Pictures shared the licensing rights for the film's characters with Marvel Studios, the licensing fees for the Blu-ray release of the film and the PS3 version of the film's video game were not shared. Those fees stayed within the Sony conglomerate. When the James Bond film *Quantum of Solace* (2008) was released on Blu-ray and for the PS3, all of the licensing fees for these products stayed within the conglomerate, and these revenues merely were transferred from one division of Sony to another.

Indeed, Sony's synergy is a prime example of the whole of the conglomerate proving to be greater than the sum of its parts. For example, Sony had lost a great deal of money on the PS3 console. Yet even with lagging sales, the PS3 helped Sony to win the hi-def DVD war, and that may prove to be the greater victory. Now that Blu-ray has become the dominant hi-def format, Sony controls an important market segment. All companies producing new products that use this new standard (including hardware, such as DVD players, game consoles, and computers, and software, such as home videos and video games) have to secure licensing rights from Sony. Sony already collects licensing fees on all hi-def DVD players and hi-def DVDs. At some time in the near future, Sony could also collect licensing fees on all video game consoles and video games. Finally, if the format becomes the standard for optical drives, Sony will collect licensing fees from every computer manufacturer that chooses to install Blu-ray drives in its machines.

Although it has won the format war, Sony is not completely in the clear when it comes to the Blu-ray format. Blu-ray could be a significant victory for Sony, or it could be another humiliating defeat. Cable television and satellite services also offer hi-def programming, and these services can give consumers content without the need for Blu-ray technology. Microsoft, which supported HD DVD, has said it will not adopt Blu-ray for its Xbox 360; given Microsoft's dominance in the PC market, its rejection of the technology may hinder Sony's efforts to position the format in the computer market.[20] Given that both the Xbox 360 and Nintendo's Wii have online capabilities, both of these consoles could offer hi-def content via online access and forgo installing Blu-ray drives altogether. Currently, video game blogs have reported rumors of a hi-def Wii in development, but these are only rumors, and Nintendo has made no mention of adding a Blu-ray drive to the Wii.[21]

At this point, there is little indication that consumers would completely abandon physical media in either the home video or video game markets. In addition, sales of Blu-ray discs increased significantly after the demise of HD DVD, indicating that consumers may still prefer physical media even for hi-def content.[22] If consumers continue to adopt Blu-ray, both Microsoft and Nintendo may have to adopt it as well in order to offer hi-def content on physical media. Nevertheless, Sony's victory in the format war may still carry significant costs. For example, its synergistic strategies have not served it well in the game console market, a market in which it is currently losing to Nintendo.

Nintendo's Blue Ocean

At the 2006 Electronic Entertainment Expo (E3), Nintendo introduced its Wii console. The Wii represented not only a significant advancement in game technology, but also a different kind of innovation. While everyone else was focused on improving video technology,

Nintendo worked on enhancing the game play of video games. It developed a wireless remote control that could track motion in three dimensions. Other game consoles like the Xbox 360 and the PS3 also had wireless remotes, but these consoles lacked this motion-tracking capability. Therefore, the Wii introduced a new gaming experience because it motivated game players to get off the couch and move.

The Wii won the 2006 E3 Best of Show award and several other industry awards. More important than this positive critical reception, however, was the response of consumers, who rushed to stores to purchase the console. During the holiday shopping season of 2006, the Wii was the must-have gift in the U.S. market, and stores sold out their stock as soon as new shipments of the console arrived. Moreover, while video games have generally targeted a market segment of young males, Nintendo made a strategic effort to move beyond the traditional core market for the Wii. In fact, Nintendo has been vocal about its new "blue ocean strategy." Perrin Kaplan, a vice president for marketing and corporate affairs for Nintendo of America, offered this explanation in an interview with *Forbes:*

> Seeing a Blue Ocean is the notion of creating a market where there initially was none—going out where nobody has yet gone. Red Ocean is what our competitors do—heated competition where sales are finite and the product is fairly predictable. We're making games that are expanding our base of consumers in Japan and America. Yes, those who've always played games are still playing, but we've got people who've never played to start loving it with titles like *Nintendogs, Animal Crossing* and *Brain Games.* These games are Blue Ocean in action.[23]

The blue ocean strategy (BOS) was introduced by W. Chan Kim and Renée Mauborgne, who distinguish between red oceans—industries with clearly defined boundaries and fully competitive markets—and blue oceans, which "denote all industries not in existence today—the unknown market space, untainted by competition."[24] The BOS is not necessarily new, and Kim and Mauborgne offer several historical examples, including the Ford Model T and Apple personal computers. In contemporary market contexts, however, Kim and Mauborgne argue that a BOS should have specific defining features: "Create uncontested market space. Make the competition irrelevant. Create and capture new demand. Break the value/cost trade off. Align the whole system of a company's activities in pursuit of differentiation and low cost."[25] With its BOS, Nintendo is looking beyond the target market of hardcore gamers and reaching out to other demographic groups to expand its consumer base. This is a significant departure for a company that has historically targeted a young male market.

Nintendo was founded in 1889 as a company that produced handmade *hanafuda* (literally, "flower cards"), playing cards painted with various symbols that are often used in a simple matching game. During the 1970s, Nintendo expanded its operations and began manufacturing toys, including light-beam and laser shooting games that were precursors to its video games. During this period, the company entered the video game industry indirectly by licensing and manufacturing Magnavox's Odyssey console for the Japanese market. In 1977, Nintendo released its own branded video game system in Japan, and in 1985 the company released its Nintendo Entertainment System (NES) in the U.S. market.[26]

Unfortunately, the U.S. video game market was still in bad shape in 1985, having pretty much collapsed two years earlier due to missteps by major companies like Atari, which had released the unsuccessful *ET* video game.[27] Therefore, retailers were extremely reluctant to even stock the game console, or any other video game system for that matter. Given this reluctance, Nintendo decided to focus on the New York market during the Christmas season of 1985, and it tried to place the NES in Toys R Us, Sears, Circuit City, and Macy's. Still, the NES met with resistance, and Nintendo responded with extraordinary concessions. As David Sheff explains:

> [Nintendo agreed to] stock the stores and set up displays and windows. Nobody had to pay for anything for ninety days. After that period, stores would pay Nintendo for what they had sold and could return the rest. It was an offer store buyers couldn't refuse, although it was still greeted with skepticism. Then, one by one, companies agreed.[28]

The strategy worked, the NES became a successful console, and Nintendo went on to dominate the video game market.

Nintendo maintained its dominance in the video game market for several years by targeting its products to the specific core demographic of boys aged 8–14. Nintendo's targeting strategy involved three main tactics. First, the company conducted extensive, ongoing, pre-market testing of its games. As many as 1,200 children a week played Nintendo's games in testing labs and evaluated the games on a variety of criteria.[29] This information was used to tweak and correct problems in the tested games and was also aggregated and used to refigure the design process for future games.

As a second strategy, the company began publishing the magazine *Nintendo Power* in 1988. As the name implies, the magazine featured games for Nintendo's systems and was sold to the owners of those systems. As Kline et al. report, "Subscribers paid fifteen dollars a year, a price that covered most cost[s]. Any advertising revenues from Nintendo licenses were almost pure profit."[30] Finally, Nintendo set up call centers that allowed players to obtain help if their systems were not operating properly, or if they needed advice on how to play the games. Kline et al. explain the significance of these efforts:

> Not only did Nintendo make money by charging for help calls but also, and more importantly, the phone-lines, along with the magazines and other promotional activities, built a game culture. . . . But that was only part of the story. The other part was that the magazines and phone tips that grew out of the game culture enabled Nintendo to gather information about consumers and then incorporate it into its development process.[31]

Nintendo focused on its target market with precision, using these three marketing tactics to continually recalibrate its products to please the 8–14 male demographic. As a result of this sharp focus, Nintendo dominated the video game industry, later extending its dominance to the handheld video game market when it introduced the Game Boy system in 1989.[32]

Nintendo's dominance, however, would soon be challenged by competitors who entered the video game market. Kline et al. provide a detailed history of this competition, and while a discussion of many of those details is beyond the scope of this chapter, an important point emerges from this history.[33] As new companies entered the video game console market, they expanded the market to include an older demographic. For example, Sega was the first company to challenge Nintendo, and it did so by targeting males between the ages of 15 and 17. Sega appealed to this older demographic by adopting a more aggressive and less family-friendly approach in its marketing campaigns. Kline et al. note that Sega's strategy "made advertising video games entertaining, and they turned the problem of media-savvy youth into its own solution through an ad style based on irony and twisted humor, often at the expense of its rival, Nintendo."[34]

When Sony entered the market in the mid-1990s with the PlayStation console, it, too, aimed at an older and wider market, targeting males between 12 and 24. Sony believed, quite accurately, that it had strong brand recognition among post–baby boomers, who had grown up with such products as the Sony Walkman.[35] When Microsoft introduced the Xbox in 2000, it further expanded the video game console market by targeting males between 18 and 34.[36] Microsoft appealed to this market segment by positioning the Xbox as much more technologically advanced than other systems, and it offered more violent games, such as the first-person shooter *Halo.*

Though Sega initiated the strategy of targeting an older demographic, it ultimately failed in the game console market. In the twenty-first century, Sega no longer produces game consoles and has limited its activities to the game software market. Sony and Microsoft, on the other hand, have used the strategy successfully and have shifted the market so that the majority of game console owners are older than eighteen.[37] Unfortunately for Nintendo, it continued to focus on its original eight-to-fourteen market segment and soon found itself outflanked by the competition. When Nintendo introduced the GameCube console, it competed in the market with Sony's PS2 and Microsoft's Xbox. The GameCube sold only 21 million units worldwide, while the other consoles sold 122 million and 24 million, respectively.[38] Obviously, Nintendo no longer dominated the market. Instead of fighting to win back this market, however, Nintendo decided to create a new market for its products.

In marketing the Wii, Nintendo decided to seek out new consumers by appealing to those who were often overlooked by the video game industry, specifically women and older adults. As Nintendo president Satoru Iwata pointed out, "As we've stated before, we're not thinking about fighting Sony, but about how many people we can get to play games. The thing we're thinking about most is not portable systems, consoles, and so forth, but that we want to get new people playing games."[39] These marketing efforts are quite apparent in the advertisements and trailers for the Wii.[40]

One of the first trailers for the Wii featured a pair of Japanese men traveling the countryside, knocking on doors, bowing, and offering the wireless remote control while saying, "Wii would like to play." The first home they visit is in a suburban setting and houses a white, middle-class family of four. The men demonstrate the Wii and then the family joins in the fun, but it is the mother who is featured most prominently as she bowls and jogs in various Wii games. The next house they visit belongs to a family that is perhaps Latino, and while the other family members observe, the Japanese men teach the father (a greying, portly man) how to play baseball with the Wii. Another scene shows a young man in the family (possibly the son) playing a boxing game with an older woman (possibly the mother), who wins the match and jumps around in celebratory victory.

The men then pull up to a high-rise building, where a young African American man answers the door to an apartment. He and his two friends, younger adults who would normally fall within the target demographic for video games, play with the Wii. The last house they visit is out in the country, and they approach two bearded men working on a broken-down pickup truck in the yard and again make the offer, "Wii would like to play." After playing several different games, these rather rough men show their appreciation by offering a chicken in return, an offer that is comically declined. The Wii men then return to the city and offer the Wii to an auditorium full of cheering people.

The commercial depicts a diverse array of people, and it clearly reflects the strategy articulated by Iwata to entice more people to play Nintendo's games. It is important to note how many of these people fall outside the demographic usually associated with video games. The parents in the first two households seem to dominate the game play; not only do these scenes depict middle-aged women playing the video games, but the second scene shows the woman beating the younger man at a video game version of the decidedly masculine sport of boxing. This focus on women carries over to the shorter, 30-second TV spot, which also puts the mother at the center of the action. Moreover, the ads invite participation by a variety of consumers, and the phrase "Wii would like to play" literally offers the invitation. Indeed, the phrase functions as a "compelling tagline," a characteristic that Kim and Mauborgne recommend as part of a good BOS.[41]

In order to differentiate the Wii from its competition, Nintendo decided to focus on the gaming experience offered by its wireless remote. The commercials illustrate this decision, which has also resulted in creating applications unlike those typically associated with the sedentary practice of video game play. Specifically, Nintendo marketed the Wii as a health and fitness device and did so in an aggressive manner unprecedented in the video game industry. It could be argued that some previous dance games, such as *Dance Dance Revolution,* allowed players to exercise, but these games were not marketed in this way. Nintendo's Wii Fit, however, has been marketed as an exercise system.

Introduced about a year after the release of the Wii, the Wii Fit included a balance board and several health and exercise programs. Although the ad campaign for the Wii Fit represented both genders, women figured more prominently in the campaign. For example, a Wii Fit TV spot includes four women and only three men. This difference may not seem significant, but given that males make up 72 percent of the market for game consoles, the difference is striking.[42] The focus on women carried over to both the packaging and print promotion of the Wii Fit. For example, the box art for the Japanese release of the application featured a silhouette of a woman holding a yoga position. An original ad image that featured both men and women was reconfigured for a Best Buy circular so that one of the women is more prominently featured. Finally, a promotion poster shows only a woman, with the tagline "How will it move you?" It is important to note that the Wii Fit was an add-on to the original console, so anyone who purchased this add-on had to either own or purchase the console as well. Therefore, the Wii Fit campaign was actually about adding women to the target market for the Wii console.

In addition, the Wii has been embraced by medical professionals as a means of administering physical therapy to the elderly and the injured. Because most Wii games require movement, many hospitals and nursing homes have incorporated the consoles into their rehabilitation regimes. For example, the MedCentral/Mansfield Hospital in Ohio uses the console in physical therapy to assist stroke victims.[43] The Wii sports program includes activities with which many older adults are familiar, such as golf and bowling, and this familiarity makes the console more accessible to a population unfamiliar with video games. The Weill Cornell Medical College uses the Wii in its burn center to encourage patients to exercise.[44] In addition, the Medical College of Georgia received a grant from the National Parkinson's Foundation to study the use of the console to increase motor function among patients with the disease.[45] If these uses prove successful, the Wii will have created demand in the medical services industry. Considering how often video games have been maligned as the antithesis of physical activity, these uses reflect a significant departure from standard industry marketing practices.

Perhaps unsurprisingly, Sony does not seem to appreciate what Nintendo is trying to do. For example, Sony CEO Howard Stringer offered the following assessment of the Wii: "I've played a Nintendo Wii. . . . I don't see it as a competitor. It's more of an expensive niche game device."[46] If the Wii has become irrelevant to Sony, then perhaps the PS3 has become irrelevant to Nintendo as well, as recent sales figures suggest. Don Reisinger, writing for *CNET News,* observes:

> Remember Sony and Microsoft? No? Well, neither does the average Amazon customer. According to a release sent out by the online retail giant Friday, the Nintendo Wii and all its accessories dominated video game sales during the holiday shopping rush and not one mention was made of Sony's PlayStation 3 or Microsoft's Xbox 360. . . . What about all its competitors? Have they somehow entered the realm of irrelevance? I'm starting to wonder if they have.[47]

It is important to distinguish between Reisinger's use of the word *irrelevance* and Kim and Mauborgne's meaning. Reisinger seems to suggest that the PS3 and Xbox 360 may no longer matter in the video gaming market, but the Amazon figures he references do not justify that conclusion. What the Amazon figures may indicate, however, is a change in the aggregate video game market. In other words, Nintendo has reached out to new customers and created new demand in markets where neither Sony nor Microsoft seem to be competing.

What is most interesting about Nintendo's BOS, however, is how it required the company to abandon accepted industry wisdom. Typically, advances in video technology and the resulting improvements in image quality have been viewed as ultimately driving innovation in the video game industry. This certainly was the wisdom that led to the development of the PS3 with its Blu-ray drive. The Wii, on the other hand, offers no hi-def content and includes only a standard DVD drive, which is not enabled to play DVD movies. While this was an improvement over the GameCube (which still used CDs), given that the earlier Xbox and PS2 already had DVD drives, this technological "update" might have seemed hopelessly retrograde. At the time of its release in late 2006, however, the Wii enjoyed a $249 price, which was about half the cost of the cheapest PS3. One reason that Nintendo could introduce the console at such a reasonable price was that it did not incorporate expensive high-capacity processors or hi-def optical drives. Instead, Nintendo's Wii offered technology that enhanced the gaming experience, but did so without significantly increasing the price of its product.[48]

Perhaps the best indication of Nintendo's commitment to a BOS is the company's 2008 annual report. The report opens with a statement from Iwata:

> Nintendo has focused its basic strategy on expanding the worldwide gaming audience. To achieve this, Nintendo is encouraging as many people as possible around the world, through its unique hardware and software offerings, to experience and enjoy video game entertainment, regardless of their age, gender, language, cultural background or gaming experience.[49]

This is a clear and unequivocal commitment to Nintendo's BOS, and as Iwata points out, it is a strategy that has been embraced by both its hardware and software divisions. If that statement itself is not enough, the design of the report also reflects Nintendo's BOS. The first five pages of the report contain ten photos that feature thirteen women and ten men, including middle-aged parents playing the Wii with their children and an older couple playing the console. Again, young men are not excluded from these images, but the design of the report clearly reflects Nintendo's desire to expand its market and align its practices accordingly.

Nintendo has been cagey about discussing the success of its BOS in concrete terms. While the annual report visually reflects its expanded market, the report never specifically mentions female players or older consumers; instead, it contains abstract statements like the one offered by Iwata. Ultimately, however, the success of the BOS can be measured in the aggregate numbers. As of December 2009, Nintendo enjoys a 48.5 percent share in the console market and a 68.5 percent share in the handheld market. In other words, Nintendo's purpose was to get more people to buy and play video games, and these numbers indicate that more people are buying Nintendo's games.

The success of the Wii illustrates that, while technology can drive the video game market, advancements should not be limited to improving the visual graphics of games. The Wii advances technology not by enhancing the graphics but by changing the game interface and the way video games are physically played. By way of contrast, the sales of Sony's PS3 (27.8 million units and 23.1 percent market share) illustrate that the convergence of its film and video game divisions was not an advantage in this latest console war. Its synergistic strategy to win the hi-def DVD format war may have left it vulnerable to the competition offered by the Wii.

In addition, now that Nintendo has created a video game market for women, it may have created an opportunity to repurpose additional content. For example, the animated feature *Coraline* (2009), which has a young female protagonist, had video games released for the Wii, the Nintendo DS, and the PS2. Sales of the game on all platforms were underwhelming, but the choice of platforms reveals that film producers may already be thinking about exploiting the new market that Nintendo has created.[50] The Wii's success, however, has had everything to do with its BOS and little to do with the convergence practiced by Sony.

Intellectual Property and MMORPG Culture

EverQuest was one of the first 3-D massively multiplayer online role-playing games (MMORPGs), an online manifestation of the RPG (role-playing game) genre discussed in chapter 5. Sony Online Entertainment (SOE) opened the game to players in 1999, the same year the internet backbone—the main transmission lines for the internet—was upgraded to 2.5 gigabits per second. This was not mere coincidence; although some MMORPGs can be played on dial-up internet service, many recommend broadband service.[51] Consequently, the emergence of MMORPGs has closely followed the commercial deployment of broadband internet technology. As cable companies and telecommunications providers expanded their fiber optic networks, they began offering commercial broadband service in limited markets as early as 1997. Fiber optic networks are able to accommodate large amounts of data and transmit those data at faster speeds than the older telecommunications infrastructure.[52] As more neighborhoods were connected to these fiber optic networks, more households had access to broadband. Each year, more households began adopting broadband service, and by 2004, 31 percent of U.S. households had broadband.[53] That year was also when *World of Warcraft* (*WoW*) was released.

Online games have received a great deal of attention, due in part to the success of *WoW*. *World of Warcraft*, owned by Activision/Blizzard, as of 2008 had over 11 million subscribers, and the latest expansion software for the game, *Wrath of the Lich King*, "sold more than 2.8 million copies, making it the fastest-selling PC game of all time."[54] In addition, the game has become somewhat of a cultural phenomenon, and conventions are held for players with admission tickets priced at $125 each.[55] Given its share (62.2 percent), *WoW* dominates the MMORPG market; the second most popular game, *RuneScape*, has only a 7.5 percent share.[56]

In spite of this dominance, some studios have decided that the online market is a viable place to extend their franchises. For example, Disney has a MMORPG based on the *Pirates of the Caribbean* films. Players create their own pirate avatars and complete quests offered to them by Jack Sparrow, Will Turner, and Elizabeth Swann. SOE also has MMORPGs based on *The Matrix* and *Star Wars* franchises. A *Star Trek* online game is also in the works, to be licensed by CBS television studios, which separated from Paramount film studios when Viacom was split into two separate companies in December 2005.[57] Therefore, this game will not be based on the most recent *Star Trek* (2009) film, although it will still be an extension of the franchise.

As opposed to most video games, which generate revenue through sales of the game software, online video games are subscription services and maintain a relatively steady stream of revenue. Many MMORPGs do charge for the game software—*The Matrix*

Online game is one example—though others, like Disney's *Pirates of the Caribbean* game, do not. In either case, the main source of revenue from these games comes from the subscription fees, which may be charged on a monthly, quarterly, or annual basis. Often, players receive a discount if they sign up for longer subscription periods, so players who sign up for a full-year subscription will pay less per month for the service than those who pay on a quarterly or monthly basis. While this pricing model is advantageous for the player, it also benefits the game companies and film studios because it generates greater revenue up-front.

This subscription model is an attractive option for companies because revenue can be generated continuously as new players sign up for the service and current players renew their subscriptions. There can be significant turnover in MMORPG subscriptions, with new players joining and others canceling the service. The terms of these games do not allow players to be reimbursed for canceled service; instead, the service is discontinued at the end of the period for which the player has paid. In other words, once players have paid for a year of service, they receive that year of service even if they cancel their accounts before the year is over. Therefore, MMORPGs are still a more stable and predictable source of revenue than retail sales.[58] As I noted in the first chapter, the video game market can be unpredictable, and most video games lose money when they are released. MMORPGs allow companies to build up a game over time, and the subscription fees can generate a long-term revenue stream. It is easy to see why these games might be attractive to film studios. What is less clear, however, is the degree to which these studios understand or respect the cultures that develop within these games.

MMORPGs are highly social game environments. To advance in these games, players must complete several tasks (quests), some of which require them to work cooperatively with other players. In fact, established groups (guilds) have formed to support these group quests. T. I. Taylor, who has studied the culture of MMORPGs extensively, found that the social relationships that form in these games are important to the players and can be both rewarding and enduring.[59] Consequently, MMORPG players are very engaged in these games, because they provide for many players both a social network and a culture. While the game design of most MMORPGs facilitates the formation of guilds, players must initiate these guilds; therefore, the players are primarily responsible for the social networks and cultures that emerge. MMORPG players thus have a sense of ownership as a result of their personal and economic investment in the games they play.

In addition, players may spend hours developing their game avatars, earning special skills and characteristics. Players also earn special items, such as armor and game currency ("gold"), when they complete quests and defeat their enemies. In other words, MMORPGs have their own internal economies, and players work within these economies to build their avatars.

Some of this work may resemble traditional labor. Edward Castronova and his colleagues have analyzed the economies of MMORPGs and worked with other game designers to actually build one of these economies.[60] He argues that there is little to distinguish these virtual economies from material ones, and game play is in several respects like other forms of productive work. Therefore, the players of MMORPGs may feel a sense of ownership, because they have worked to earn their game property. As Castronova reminds us, economies define the nature of social relationships, and this is certainly true of the economies found in MMORPGs.[61] Guilds often maintain banks in which members can store their game currency and other valuable items. Guild members may also exchange items and help each other out with gifts and loans of game currency.

Castronova has also noted that virtual economies in MMORPGs may spill over into real world economies.[62] A grey market has emerged that allows people to purchase game currency and items using real world currency. This grey market is supported through the practice of gold farming. Gold farming occurs in "virtual sweatshops" located in countries such as China, Romania, and Mexico, where players are paid a nominal wage to cultivate currency and other game resources that are then sold online to players in developed countries, mainly the United States.[63] Thus, while some players labor in the game to generate the necessary resources to advance, others pay someone else to labor for them. Ultimately, it is often the brokers of these resources, not the players who labor to produce them, who benefit the most. In 2009, a farming website was sold for $10 million in real world currency. Mike Fahey, writing on the game blog *Kotaku,* asked, "I wonder how much of that $10 million the farmers themselves will see?"[64]

The practice of gold farming is controversial within gaming communities and violates the end user licensing agreements (EULAs) of many MMORPGs, a set of rules enforced by the company that developes the game and maintains the game servers. Where gold farming is concerned, however, enforcement can be uneven and not as aggressive as some game players would like. Therefore, some players have taken enforcement into their own hands like virtual vigilantes: hunting down, harassing, or even killing the avatars of game players they suspect of farming.

Unfortunately, these enforcement tactics often take on a racist quality. Although gold farmers reside in several different countries, many players assume that most gold farmers are Chinese, and the phrase "Chinese gold farmer" has become all too common. Lisa Nakamura has analyzed the practice of gold farming and the racism that has been directed at Chinese players.[65] She observes that some players have taken their campaign against gold farming outside of the game, creating *machinima* protesting the practice. Machinima are user-created films that are recorded in the game environment and are often posted online. Nakamura discusses one machinima produced in *WoW* that mocks the Chinese as cheap labor and depicts and celebrates the killing of a supposed Chinese gold farmer.

Ironically, the production of machinima often violates the same EULA agreement that these protesting players want to see enforced. When it comes to user-created content, some companies have been very active in enforcing EULAs. Taylor relates the story of a player of *EverQuest* who posted a piece of fan fiction depicting the rape of a 14-year-old girl within the game world.[66] Because the story was published under the player's avatar name, SOE closed the player's account. This was a controversial move, particularly given that the story was posted on a web page that was not owned by SOE. Initially, SOE claimed that the story was detrimental to the game and would alienate other players, and its actions were fully within its intellectual property rights. Ultimately, SOE apologized for its actions and admitted that it may have overreached where the practices of fan fiction are concerned. Yet, as Taylor notes, the incident reveals the complicated relationship players have with the companies whose games they play.[67]

In their book *The State of Play: Law, Games and Virtual Worlds*, editors Jack M. Balkin and Beth Simone Noveck juxtapose two essays, one by Julian Dibbell and the other by James Grimmelmann.[68] Although the essays present both sides of the EULA issue, with Dibbell seeming to side with game designers and Gimmelmann with game players, both authors reach a similar conclusion. They find that, in the context of MMORPGs, EULAs are not just legal documents; they also function as social and political contracts. Given that MMORPGs generate strong social communities, this conclusion makes sense, because EULAs control game play and game play is the basis of the interactions that form these communities.

Yet, to view EULAs as social and political contracts requires an understanding of and respect for the unique characteristics of online games, characteristics that distinguish them from other forms of licensed intellectual property. It is not clear that the lawyers who handle the intellectual property rights of major film studios either understand or respect the distinct nature of MMORPGs. In fact, where Disney is concerned, the company clearly does not. When players sign up for Disney's *Pirates of the Caribbean* online game, they must not only affirm the "Member Agreement" for the game but also the "Terms of Use" agreement that covers all of Disney's online content. In this way, Disney has operationalized its EULA so that its online game is *not* treated as a distinct intellectual property.

While Disney's "Member Agreement" is unique to the game and seems to acknowledge the unique characteristics of MMORPGs, it clearly does not respect any sense of ownership that players may have:

> You do not own the Account, nor do you own any data stored on our servers. When using the Service, you may accumulate things, including currency, treasures, skills, equipment, and other items, that reside as data on our servers. All Account data may be deleted, altered, moved, or transferred at any time for any reason or no reason at all in Disney's sole and absolute discretion. In the event that such data is [*sic*] corrupted, destroyed, or otherwise lost, you acknowledge that Disney shall not be subject to any liability.[69]

In all fairness, other MMORPGs, including *WoW*, have similar stipulations about player accounts. But *WoW*'s EULA differs in that it allows for the presence of third-party licensed property in the game. In other words, Activision/Blizzard offers players more latitude to develop their own game-based content. Furthermore, Activision/Blizzard has sometimes been lenient when it comes to enforcing the *WoW* EULA, hence the practice of gold farming and the presence of player-created machinima on the web.

Disney has a reputation for aggressively pursuing intellectual property violations; it is thus doubtful that it would ever be as lenient as Activision/Blizzard, and for good reason.[70] In most MMORPGs, the game is the primary product, and while these games can be franchises in and of themselves (*WoW*, for example, has spawned a variety of ancillary products), the success of the franchise is dependent on players' satisfaction with the game. Activision/Blizzard must be willing to accommodate the players of *WoW* because the MMORPG is its main source of revenue.

Where film spin-offs are concerned, however, the film is the primary product and the game is ancillary, so the success of the franchise is not dependent on game players' satisfaction. For example, as of April 2008, Disney's *Pirates of the Caribbean* had only 10,000 active subscribers; as of May 2006, *The Matrix Online* had 30,000; and as of October 2007, *Star Wars Galaxies* had 100,000.[71] In all of these cases, the subscription numbers indicate that the games do not even come close to generating the billions of dollars the associated films made in box office receipts alone.

Furthermore, a company like Disney has a brand identity that is far different from that of Activision/Blizzard, or even Sony. The use of *WoW* intellectual property by players to create racist texts seems not to have diminished the game brand. The fact that Sony apologized to the player who created the fan fiction mentioned earlier also seems not to have generated much negative publicity; on the contrary, the decision to apologize was probably informed by the negative reaction from the players of *EverQuest*. On the other hand Disney has several media products marketed to young girls, including the Hannah Montana, *High School Musical*, and Jonas Brothers franchises. Given the nature of the Disney brand, it could not afford to be as forgiving if its intellectual property had been used to depict the rape of a fourteen-year-old girl. If Disney were forgiving of such a use, the reaction from parents would be quite negative. This reaction would, in my mind, be justified, and when licensed content is appropriated to depict either racial or sexual violence, a company may have both the right and the moral obligation to censor such appropriation. When media conglomerates exercise their rights, however, the motive is often financial rather than moral.

If *WoW* sets the standard for MMORPG success, and if Activision/Blizzard is the exemplar for cultivating and accommodating player culture, then online games may have a limited future where the film spin-off is concerned. In other words, where convergence has been accomplished by similar business interests and market conditions in the video game and film industries, the unique nature of MMORPGs may mark a limit point. In the practice of product licensing, the control of intellectual property is important to the film studios. This need to maintain and control their intellectual property may not allow studios to accommodate the sense of ownership that many MMORPG players have developed in regard to online games. Finally, as I noted earlier, MMORPG film spin-offs have not enjoyed a great deal of success: SOE, for example, ended *The Matrix Online* game in July 2009.

For convergence to work with MMORPGs, companies would have to be selective about the film content they repurpose in online games, and they would have to allow players more latitude in the use of that content. In addition, repurposing film content may not be the best way for some media conglomerates to utilize MMORPGs. Instead, they might use these games as a way of creating demand for other technologies. For example, Sony has introduced "Home," a virtual environment that uses PS3's network capabilities and is designed to compete with the popular *Second Life* online environment. Time Warner, one of the largest providers of cable and broadband internet service in the United States, could develop MMORPGs as a way of promoting its services rather than as a means of promoting its films. In either case, MMORPGs would augment the commercial viability of other technologies and services offered by these companies. Yet, given the performance of current film spin-off MMORPGs, it would appear that film studios are not very successful in repurposing their films for this particular genre.

Let's Play War

The final issue I will discuss involves the development of technology through the collaborative efforts of the film and video game industries and the military. There is a long history of collaboration between the military and the film industry, particularly where the development of technology is concerned, dating as far back as the Civil War.[72] In World War I, this collaboration took an ideological turn when William A. Brady, president of the National Association of the Motion Picture Industry and the World Film Corporation, approved the Four Minute Men program. Under this program, speakers delivered orations promoting the war or recited patriotic poems during the intermissions at motion picture theaters throughout the United States.[73]

This type of patriotic promotion reemerged during World War II with such films as the rather campy *Stage Door Canteen* (1943). Supposedly set in the actual Stage Door Canteen in New York's theater district, but actually shot on an RKO sound stage, the film depicts several stars entertaining the troops before they are shipped off to war. In the years following the war, the World War II film became a ubiquitous subgenre, and in my household my father insisted on watching any and all that were broadcast on network television. In the post-Vietnam period, the patriotic zeal of war films began to decline, but it continues to bubble up even in film's newer manifestations, such as music videos. The band Three Doors Down, for example, produced an intensely patriotic video for its song *Citizen/Soldier,* which is shown as part of the previews in some film theaters. The video, punctuated with patriotic imagery, also has functioned as an advertisement for the National Guard.

The video game industry also shares a long history of collaboration with the military, and some early games and gaming systems were developed by people working in conjunction with the U.S. Defense Department. For example, Ralph Baer was working on radar systems for the navy when he came up with the idea of a home video game system; that idea materialized in the form of the Magnavox Odyssey, one of the first video game consoles.[74] Over the years, the video game industry became even more aligned with the military. In the early 1980s, the military recognized video gaming as a means of recruitment, and even today the National Guard web page has a "Prism Guard Shield" game that can be downloaded (http://www.nationalguard.com/events/prism-guard-shield-game). Video games have also become integral to the training of military personnel, and some training software has been repurposed for the commercial market. For example, *Spearhead* is a tank simulation game that was originally developed for actual military training. In addition to software, video game hardware technology has also been incorporated into military designs. The PS2 control pod, for example, served as the design template for the interface of remote-controlled military equipment and guided missile systems. Because many new soldiers were already familiar with this control pod, their training time on the military equipment could be significantly reduced.[75]

This collaboration has now manifested as a convergence of the film industry, the video game industry, and the military industrial complex. James Der Derian refers to this convergence as the "military-industrial-media-entertainment network," or MIME-NET.[76] The MIME-NET is exemplified by the Institute for Creative Technologies (ICT) at the University of Southern California, a project funded by the army that brings together scholars and members of the entertainment industry to design new technologies in interactive media and virtual simulations. The home page for the institute reveals exactly how these new technologies might be applied: pictures flash on the page showing virtually rendered tanks and army equipment, a solider in the field in a virtual reality helmet with a control pod in his hand, and a virtual character called "Sgt. Blackwell."[77] One of the images portrays a simulation program for language acquisition and virtual dialogue, and it depicts two men who appear to be vaguely Middle Eastern. Given the army's current military engagements, it is easy to imagine how this simulation program might be used to train soldiers, and the contexts and kinds of people and conversations it might be used to simulate.

The ICT is hardly unique. As Mia Consalvo and Toby Miller observe, numerous other military-funded projects involve the film and video game industries; they identify examples of such programs at Carnegie Mellon University and the University of Central Florida.[78] These programs represent a high level of collaboration between the military and the entertainment industry, one that inevitably raises some challenging questions about the purpose and use of the technologies they develop. When the ICT had its 1999 inaugural press conference, however, few of these critical questions were raised. Der Derian attended this opening ceremony and did succeed in questioning Jack Valenti, then president of the Motion Picture Association of America, about the possibility that the institute would further blur the line between entertainment and propaganda.

Valenti's response, according to Der Derian's account, was dismissive and condescending. Der Derian concludes, "Valenti and his cohorts at the ICT seemed unaware of their own potential role in the . . . aestheticizing of violence, the sanitization of war."[79]

On this point, Der Derian's complaint begins to resemble the thesis of the media effects research on video game violence. Fortunately, he does not continue down this road, instead expanding his critique into a much broader assessment of the ideological impact of the MIME-NET. He argues that the convergence of the military and the entertainment industries, and the continual relaying of technology between military applications and commercial entertainment, erase the distinction between times of war and times of peace, creating a permanent state of "interwar."[80] In this state, the critical engagement of any one war and the policies that inform it become much more difficult. After all, if war is understood as a permanent state or becomes a given in the public consciousness, then the presumption is always for war and never for peace.

Nick Turse draws a similar conclusion in his analysis of what he calls "The Complex."[81] Turse argues that the military has moved beyond the traditional defense manufacturing industries and now has alliances with the media and entertainment industries. He maintains that these new alliances allow the military to infiltrate the lives of civilians in multiple ways. The outcome, as Turse describes it, is a realization of the conditions Dwight Eisenhower once warned against:

> From the 1940s through his years in the White House, Eisenhower repeatedly decried unrestrained defense spending as a pathway to a "garrison state" where the military would hold extraordinary sway over the nation. Today, the United States has begun to resemble what Eisenhower feared. Having garrisoned the globe, the Complex is returning home in new and unnerving ways.[82]

Roger Stahl illustrates how this threat is manifested in his analysis of war-themed video games. Stahl discovers in these games the emergence of a "citizen-soldier" identity, "produced by the changing configurations of electronic media, social institutions and world events."[83] He maintains that this new identity erases the distinction between the political citizen and the apolitical soldier and ultimately depoliticizes the public sphere. Stahl concludes, "[T]he virtual citizen-soldier's integration into a sanitized fantasy of war is a seduction whose pleasures are felt at the expense of the capacity for critical engagement in matters of military might."[84] The Three Doors Down video I mentioned earlier suggests that the National Guard has already adopted the term citizen-solider in its marketing campaigns.

While Turse's Complex and Der Derian's MIME-NET both illustrate ways that the film and video game industries have converged with the military, neither of these scholars have exhausted this area of study. The United States is currently engaged in an ongoing "war on terror," and even under the Obama administration there has been little indication that these military efforts will end any time soon. Even if the United States discontinued its current military engagements, that would not necessarily bring an end to projects like the ICT or the convergence they signify. In other words, like the war on terror, the alliance among the military, the film industry, and the video game industry is ongoing, and scholarly analyses of this alliance should continue as well.

It is important to recognize, however, that the motives behind the type of convergence Turse and Der Derian describe are quite different from those outlined in this book. Although projects like the ICT are designed to create new technologies, the primary application of these technologies is informed by a political agenda. Therefore, film and video game convergence in the context of these military projects requires a different analysis than I have provided. Whereas the convergence I have considered is specific to the advantages enjoyed by both film studios and game developers, the convergence that Turse and Der Derian discuss has much broader ideological implications. Textual analyses of specific games might be useful to investigate this type of convergence, but any critical analysis would need to be anchored in those broader implications and not in the specific industrial conditions I have discussed in the preceding chapters. A broader ideological critique should drive future studies in this area, and the analyses should include other forms of media in addition to film and video games.

In this chapter, I have focused on technology and its relationship to film and video game convergence. In the case of the PS3, convergence has not been a benefit, at least where the game console market is concerned. In addition, film studios have seen limited success in adapting their content for MMORPGs. Therefore, there seem to be some limits to the practice of convergence. Programs like the ICT, however, indicate that convergence may continue in the future, but in ways that may have some negative implications.

In a May 2009 interview in *Wired* magazine, *Hobbit* director Guillermo del Toro expressed a much more optimistic view of the future of convergence:

> In the next 10 years, we're going to see all the forms of entertainment—film, television, video, games, and print—melding into a single-platform "story engine." . . . The moment you connect creative output with a public story engine, a narrative can continue over a period of months or years. It's going to rewrite the rules of fiction.[85]

Though del Toro acknowledges his disappointment with the video games that have been released for his own films, he remains optimistic about the future possibilities, contending, "In the next 10 years, there will be an earthshaking *Citizen Kane* of games."[86] Perhaps there will, but I do not think that such a game will be the product of convergence.

As I was working on this book, I occasionally mentioned the project to other gamers. They always expressed interest, but many also expressed reservations. As one remarked, "You know, most of those games based on movies are really bad." I found it difficult to disagree. While I thought that many of the games I played and analyzed for this book were of good quality (and being an RPG fan, I particularly liked the *KH* games), as I noted in the previous chapters, some of the games were of poor quality, with repetitive game play and unchallenging missions. My analysis in this book offers a possible explanation: in the context of convergence, the design of games accommodates interests other than those of quality game play. When a film studio begins working with a game developer on a video game spin-off, the studio's interest is in the profitability of the film. Therefore, the game design often must incorporate promotional aspects that need not be included in a game designed to stand on its own.

The *LOTR* games and the games based on the Marvel Studios

films provide examples of promotional practices informing game design. These games' reward systems were designed around promotional messages and materials, and additional elements of the game design also reference other texts in the franchise. With *The Godfather* game and the *KH* games, the promotional practices are subtler but more deeply rooted. *The Godfather* game became a way for Paramount to revive the franchise even without Coppola's cooperation. The *KH* games allowed Disney to weave together the narratives of many of its films into a single grand narrative about the corporation's supposed benevolence. While these games do not have overt promotional messages, such as the film actors relating the experience of the game to the film, they nevertheless serve the interests of the film studios involved. In addition, Sony's interest in promoting the Blu-ray format determined the design of the PS3 console, and that interest may have caused the console to lose market share to Nintendo's Wii. Finally, film studios' interest in maintaining control over their intellectual property seems to be in conflict with the very gaming culture that makes MMORPGs successful.

However, the problem does not rest solely with the film studios. As del Toro observes, game developers can be extremely conservative and resistant to new ideas.[87] As I mentioned in the first chapter, game developers are committed to a set of game genres with familiar characteristics, because they know they can market those genres. In fact, one reason that film studios and game developers have been able to work together is that they both believe that their products should conform to genres that they can market. In other words, convergence does not appear to have brought about significant change in either industry; on the contrary, it has often caused existing practices to become more entrenched.

I would like to believe that the future holds in store a game that has the cultural resonance of *Citizen Kane,* but I rather doubt that such a game will be a spin-off of *Citizen Kane.* Simply put, I am not convinced that such a game will emerge from the practices of convergence described in this book. Even if the type of "public story engine" del Toro imagines gains traction with media consumers, the forms of media he identifies as fueling this engine (including television and print) are already integrated into the major media conglomerates that own film studios. Indeed, the continuing narratives

NOTES

Acknowledgments

1. Brookey and Booth 2006.

1. Playing Together

1. These box office revenue data were retrieved from http://BoxOfficeMojo.com.

2. Grover 2007, par. 5.

3. Zacharek 2007, par. 11.

4. Scott 2007, par. 6.

5. Hocking 2007, par. 4.

6. D. Jenkins 2007.

7. Thorsen 2005.

8. Kent 2001.

9. Kohler 2005.

10. Maslin 1993, par. 1.

11. These box office revenue data were retrieved from the Internet Movie Database (http://www.imdb.com).

12. Remo 2006.

13. This information was compiled by correlating data from http://BoxOfficeMojo.com with product listings from Amazon.com.

14. Lang 2008, pars. 1–3.

15. This information was compiled by correlating data from http://BoxOfficeMojo.com with product listings from Amazon.com.

16. Keighley 2007.

17. Hollywood Movies 2005.

18. Ovadia 2004.

19. Although it did not prove to be a convergence technology, the CD format was important to the development of video game technology. It better

accommodated full-motion video and was much cheaper to produce than the ROM cartridges that were the standard video game format for software. In fact, CD technology was instrumental in the development of Sony's PlayStation platform, a subject that is given more consideration in chapter 6.

20. Brookey 2007.

21. Kent 2001.

22. Kline et al. 2003.

23. Poole 2000.

24. Hunter 2000.

25. Ibid.

26. Sedman 1998.

27. Ibid.

28. Consumer Electronics Association 2005.

29. Consumer Electronics Association 2006.

30. Mitchell 2003, par. 2.

31. Snider 2003.

32. Rothman 2004.

33. Fineman 2005; Movie Gallery 2007.

34. Brightman 2007.

35. I should note that the DVD playback capability in the Wii drive has been disabled, though it is possible to hack the drive to play DVDs. Jesus Díaz (2007), writing for *Gizmodo*, suggested that Nintendo was trying to hold down costs by forgoing the licensing fees required for DVD playback capability.

36. Epstein 2005.

37. Kerr 2006.

38. Daniels et al. 1998.

39. Dekom 2004.

40. Kerr 2006.

41. Díaz 2008.

42. Aphra Kerr (2006) discusses the importance of genres to video game marketing practices, while Mark J. P. Wolf (2001) provides a more extensive explanation of the specific genres. An FPS, as the name implies, is a shooting game in which the player has a first-person point of view of the action. An RPG is a game in which players adopt game identities; I will go into more detail on this particular genre in chapter 5. These are only two examples of video game genres, and there are several others.

43. Kerr 2006.

44. Acland 2003.

45. Kline et al. 2003.

46. Surette 2006.

47. Green 2009.

48. Writers Guild 2008.

49. Interactive Contracts FAQ 2009.

50. Fritz 2008, par. 4.

51. Handel 2009, par. 4.

52. As Handel points out, sometimes this flexibility is not so much the product of cooperation as a sign of union administrative failure.

53. Ovadia 2004, 447.

54. Raugust 1995.

55. Williams 2002.

56. Kline et al. 2003.

57. Ovadia 2004.

58. Video games are also released for PCs, but Microsoft does not require a licensing fee on games developed for Windows operating systems.

59. Kerr 2006.

60. Sherry 2001; Slater et al. 2003; Smith et al. 2003.

61. H. Jenkins 2006.

62. The Serious Games Initiative is an excellent example of this area of study (www.seriousgames.org/index2.html).

63. Adorno and Horkheimer 2002.

64. Fiske 1987.

65. H. Jenkins 2006, 247.

66. Kline et al. 2003.

67. Kerr 2006.

68. Miller 2006.

69. Some scholars have chosen to use the term "digital games" in their studies. This choice is supported by the Digital Games Research Association (DiGRA) because it better encompasses the various ways that games manifest. My focus, however, is specific to convergence, and the Hollywood studios use the term "video games" when they refer to these licensed products. Therefore, I have chosen to use the term "video games" as well.

70. Turow 1997.

71. Investor Relations 2009, par. 1.

72. Satariano 2009.

73. Reisinger 2009, pars. 6–7.

74. This is particularly true given that consolidation has occurred even on the retail level of the game industry. In 2005, video game specialty retailers GameStop and EB Games merged. The new company still operates stores under both brand names, and according to its 2007 Annual Report, the company has 5,264 stores in 16 countries. The company also publishes *Game Informer* magazine, which it describes as the "largest multi-platform video game magazine in the United States based on circulation, with approximately 2.9 million subscribers" (GameStop Corp. 2008, 2). In other words, even the video game industry understands the importance of multiple media platforms.

75. Bogost 2007, 28–29.

76. Huizinga 2000, 10.

77. Aarseth 2001.

78. Wolf 2001.

79. Ibid., 93.

80. Bogost 2007, 43.

81. I was assisted in this process by my undergraduate students, who received practicum credit for playing the games while I took notes. Although I played parts of all of the games, these students played the games to the finish, or "beat" the games. I needed this assistance because I found it almost impossible to simultaneously play the games and take good notes.

82. Grieb 2002; Howells 2002; Mactavish 2002.

83. These box office revenue data were retrieved from http://BoxOfficeMojo.com.

84. Kohler 2005.

2. Playing the Games, Being the Heroes

1. The box office revenue data that appear in this chapter were retrieved from http://BoxOfficeMojo.com on September 9, 2008.

2. Electronic Arts 2004.

3. Company History 2003.

4. Throughout this book, I will use *LOTR* to refer to the *Lord of the Rings* franchise.

5. This tagline appeared in television and print ads for the game, on the EA web page, and on the packaging for the game.

6. Sales figures were retrieved from http://VGChartz.com on September 7, 2008.

7. Garite 2003, 2. I should add that Garite does not agree with this concept of interactivity, offering instead a materialist critique in which he argues that game play is work. His analysis, however, indicts game play generally and is not specific to the synergistic connections between games and films.

8. Sawyer et al. 1998, 112.

9. Filiciak 2003.

10. Rehak 2003, 111.

11. Tong and Tan 2002, 99.

12. Bryce and Rutter 2002, 76–77.

13. Fiske 1987; Morley 1980.

14. Wolf 2003.

15. *Rockstar* in Hot Coffee 2005.

16. Bryce and Rutter 2002.

17. Crawford 2003.

18. Marshall 2002.

19. Ibid., 73. Marshall illustrates his point by drawing examples from *Lara Croft: Tomb Raider* and *Final Fantasy,* but he does not actually critique these video games.

20. Ang 1996, 12.

21. Grossberg 1995, 75.
22. H. Jenkins 1992, 17.
23. Ibid., 18.
24. Ibid.
25. Documentary 1 2001.
26. Like films, video games also have trailers that are used to promote the games. These trailers are shown at trade shows, appear on web pages, and are often recut into television commercials.
27. Oddly, however, the game fails to completely respect the new characters when they are unlocked. If levels are played again with different avatars controlling the action on the screen, the cut scenes, so essential to integrating the intertextual elements of the film and the game, will not display a character at all. In fact, in at least one case—"Shelob's Lair"—if the player uses a character *other* than Sam to beat the level, the cut scene will depict the giant spider being slain without a slayer present.
28. Ott and Walter 2000, 440.
29. Tong and Tan 2002, 103.
30. Tong and Tan 2002.
31. Figures were taken from http://BoxOfficeMojo.com.
32. Actually, this lawsuit has everything to do with the synergistic practices I mentioned in the first chapter. New Line sold the rights for a book, DVDs, and other merchandise to Time Warner at below market value. These deals benefited Time Warner as a corporation, because they significantly reduced the gross revenue for the *LOTR* franchise, of which Jackson received a percentage. In other words, by selling the rights at a reduced rate, Time Warner was able to cut itself a deal, while simultaneously cutting the amount of money it paid to Jackson.
33. Del Toro to Direct 2008.
34. Graser and McNary 2008.

3. Coppola Sleeps with the Fishes

1. AFI's 100 Years 2006.
2. Fritz 2006.
3. Smith 2005, 3–4.
4. Lapsley and Westlake 1989.
5. Thorsen 2006.
6. Reiner 2006.
7. Vital Stats 2007.
8. Fritz 2006, 6.
9. Sarris 1963.
10. Schumacher 1999.
11. Ibid., 96.
12. Rosen 1974, 47.

13. Chown 1988.
14. Ibid., 60.
15. Epstein 2005.
16. Wexman 2003, 1.
17. Bart and Guber 2002.
18. Chown 1988.
19. Schumacher 1999.
20. These revenue figures were taken from http://BoxOfficeMojo.com and reflect the gross receipts for *The Godfather, The Godfather II,* and *The Godfather III.*
21. Barthes 1977.
22. Fiske 1987.
23. Sawyer et al. 1998, 112.
24. H. Jenkins 2006.
25. Bryce and Rutter 2002.
26. Kerr 2006.
27. Zimmerman 2004, 157.
28. Bart and Guber 2002.
29. For this analysis, I have chosen to compare the game to the narrative of the film, not to the narrative of Puzo's book. It is clear from the way the game was promoted and packaged that the film, not the book, is the referent for the video game.
30. Filiciak 2003; Rehak 2003.
31. Schumacher 1999.
32. Fritz 2006.
33. Block 2008; Graser and McNary 2008.

4. Marvel Goes to the Movies

1. Raviv 2004.
2. Ibid.
3. Christiansen 2004.
4. Raviv 2004, 14.
5. Raviv 2004.
6. Data retrieved on June 28, 2009, from http://www.BoxOfficeMojo.com/alltime/domestic.htm.
7. Sales figures are from the Internet Movie Database (http://www.imdb.com).
8. Marvel Entertainment 2008.
9. Finke 2008.
10. McAllister 2001.
11. H. Jenkins 1992.
12. Pustz 1999.
13. Ibid., 114.
14. Ibid., 56.

15. Raviv 2004, 67.
16. Fiske 1987.
17. Ott and Walter 2000.
18. Rowles 2006, par. 2.
19. Brookey and Westerfelhaus 2002.
20. Everett 2003, 6.
21. In this analysis, I have incorporated many references to the history of particular Marvel characters. More information about these characters can be found at http://marvel.com/universe/Main_Page.
22. Marvel Entertainment 2006.
23. Raviv 2004.
24. We were not alone in our assessment. The IGN website (http://www.ign.com) gave the game a "poor" rating with a score of 4.7 on a 10-point scale. The first *FF* game fared a little better, with a "passable" rating and a 6.5 score.
25. Box office data were retrieved on October 16, 2008, from http://www.BoxOfficeMojo.com/alltime/world.
26. Marcus 2002, 119.
27. Ibid., 136.
28. These box office revenue data were retrieved from http://BoxOfficeMojo.com.
29. Johnson 2007.
30. Raviv 2004.
31. These box office revenue data were retrieved from http://BoxOfficeMojo.com.

5. Disney Saves the World(s)

1. Sheff 1993.
2. Raz 1999.
3. Van Maanen 1992.
4. Ibid., 23.
5. Mattelart 1980; Schiller 1979.
6. Hall 1990.
7. Sinclair 2004.
8. Raz 1999.
9. Ibid., 247.
10. Ibid., 258.
11. Van Maanen 1992.
12. Wolf 2001, 130.
13. Burn and Carr 2006.
14. In fact, the *Final Fantasy X* game inspired me to begin studying video games. In his book *Trigger Happy,* Steven Poole (2000) gives an account of how, after several years of dismissing video games as adolescent toys, he was amazed to witness the visual complexity of contemporary

video game technology. I had a similar experience when I first witnessed the *Final Fantasy X* game played on a PS2. I, too, was amazed and immediately knew that I needed to change my research agenda.

15. Kent 2001.

16. Sales figures are available from http://VGChartz.com.

17. Kohler 2005.

18. Kent 2001.

19. Sales figures are available from http://VGChartz.com.

20. Kohler 2005.

21. The box office revenue data that appear in this chapter were retrieved from http://BoxOfficeMojo.com on March 19, 2009.

22. Given the methodological choices I outlined in the first chapter and the fact that these games were not available for the PS2, I did not include them in my analysis. The *Chain of Memories* game was finally released for the PS2 in December 2008.

Although the *Chain of Memories* game provides a narrative connection between *KH* and *KH II,* there are factors about its release that complicate its function as a bridging text. First, when the game was finally released on PS2, the players who adopted the game were receiving the textual information it contained out of sequence and well after the release of *KH II.* Second, sales figures for the PS2 release indicate that few of the players who bought *KH II* bought *Chain of Memories,* and therefore most players did not experience the narrative bridge contained in the game. These, however, are points about audience reception, and my analysis focuses on the interests of production and textual strategies. When it comes to textual strategies, *Chain of Memories* offers little that is new. Instead, the game offers many of the same worlds and characters that appeared in the first game, and some elements are repeated; for example, Sora again meets Cloud in the Olympus Coliseum world and must fight and defeat him once again in order to progress in the game. After carefully considering these points and playing through several levels of the game, I decided to omit it from my analysis.

23. Sales figures are available from http://VGChartz.com.

24. Capodagli and Jackson 1999.

25. Powers and Brookey 2005.

26. Catholic League 2000.

27. Hill 2004.

28. Rose 2004.

29. Gavin 2003.

30. Ahrens 2003.

31. Pixar 2003.

32. Hudson 2004.

33. Powers and Brookey 2005.

34. As mentioned earlier, a "boss" fight is a major battle in the game, one that must be won before the player can advance.

35. I will use the full name Kingdom Hearts to refer to the location in the game; the italicized *Kingdom Hearts* and the abbreviation *KH* refer to the game itself.

36. Actually, Roxas was first introduced in the *Kingdom Hearts: Chain of Memories* game for Nintendo's Game Boy Advance. The *KH II* game also incorporates several new *FinF* characters, including Rai, Seifer, and Fuu from *FinF VIII*. More notable, however, is the presence of the Gullwings, specifically the characters Yuna, Rikku, and Lulu. These characters originally appeared in the highly successful *FinF X*, and they reappeared as the main characters in a sequel to that game, *FinF X-2*. They are among the more popular characters in the *FinF* franchise, and their introduction into this game reveals Square Enix's efforts to capitalize on its franchise.

37. Although a fan web page, *Kingdom Hearts III* (http://www.kingdomhearts3.net), has been published in anticipation of a *KH III* game, and it houses several videos of supposed scenes from the game (one even depicts Mickey), there has been no official announcement of a *KH III* game on either Disney's or Square Enix's web pages. As of December 2009, the Square Enix official web page for *KH* does not mention *Kingdom Hearts III*. Oddly, Disney's official web page does not list any of the *KH* games.

38. Wolf 1999.

39. Capodagli and Jackson 1999.

6. What Shall We Play Next?

1. The figures were taken from http://VGChartz.com on May 29, 2009.

2. Griffin and Masters 1996.

3. Nathan 1999.

4. Griffin and Masters 1996.

5. The figures were taken from http://BoxOfficeMojo.com on May 29, 2009.

6. Sheff 1993.

7. Gershon and Kanayama 2002.

8. Corporate Information 2007.

9. I will use the hyphenated form "hi-def" instead of the accepted initials HD to refer to high definition. I do this to avoid confusion when I refer to the HD DVD brand.

10. These prices are in U.S. dollars and reflect the pricing offered on Amazon.com for these products on October 26, 2007.

11. Thorsen 2007.

12. Prices obtained from Amazon.com on October 26, 2007.

13. The best source of this information is http://eProductWars.com, a web page that tracks the pricing of both Blu-ray and HD DVD discs, among other products.

14. Ault 2007.

15. Mick 2007.

16. Perenson 2008.

17. Crowe 2008.

18. Ganapati 2008.

19. The licensing structure for Blu-ray can be found at http://www.blu-raydisc.info. The licensing structure for HD DVD was previously posted at http://www.dvdfllc.co.jp/hd_dvd/hd_what.html. After Toshiba discontinued the format, the information was taken down.

20. Ganapati 2008.

21. Ivan 2009.

22. Ault 2008.

23. Rosmarin 2006, 11.

24. Kim and Mauborgne 2004, 77.

25. Kim and Mauborgne 2005, 18.

26. Sheff 1993.

27. Kent 2001.

28. Sheff 1993, 165–166.

29. Kline et al. 2003.

30. Ibid., 120.

31. Ibid., 121.

32. Kent 2001.

33. Kline et al. 2003.

34. Ibid., 131.

35. Kline et al. 2003.

36. Ibid.

37. Kerr 2006.

38. These figures were taken from http://VGChartz.com.

39. Gantayat 2006, 4.

40. Trailers are longer advertisements that are shown mainly at trade shows, but can also air on television.

41. Kim and Mauborgne 2005.

42. Kerr 2006.

43. Kinton 2008.

44. Wii-habilitation 2008.

45. Hinley 2008.

46. White 2008, 12.

47. Reisinger 2008, pars. 1–3.

48. Technology business blogger Michael Olenick used his "BOS Createware" to map out a "strategy canvas" for the Wii. His analysis, which highlights some of the issues I have raised here, can be found at http://www.valueinnovation.net/2008/04/nintendo-wii-blue-ocean-strategy.html.

49. Iwata 2008, 7.

50. Sales data can be found at http://VGChartz.com.

51. For example, the system requirements for *EverQuest II* and *Star Wars Galaxies* include a 56K internet connection, but they recommend broadband. The system requirements for *World of Warcraft* indicate that the

game can be played with a 56K connection, but I have always played the game through a broadband service.

52. Dodd 2005.

53. U.S. Residential Broadband 2007.

54. *World of Warcraft* Subscriber Base 2008.

55. Blizzard claims that BlizzCon is for players of all of its games, including *StarCraft* and *Diablo,* but Blizzard has not released updates for these two games in years. Although the company is currently working on both *StarCraft II* and *Diablo III,* as of June 15, 2009, there was no scheduled release date for either game. Most likely, many of the attendees at this convention are drawn by the appeal of *WoW.*

56. Woodcock 2008.

57. The history of CBS and its relationship with Viacom can be found at www.cbscorporation.com.

58. Ducheneaut et al. 2006.

59. Taylor 2006.

60. Castronova et al. 2009.

61. Castronova 2005.

62. Ibid.

63. Dibbell 2006.

64. Fahey 2009, par. 5.

65. Nakamura 2009.

66. Taylor 2006.

67. Ibid.

68. Dibbell 2006; Grimmelmann 2006.

69. Member Agreement 2009, par. 13.

70. Capodagli and Jackson 1999.

71. All figures have been taken from http://MMOGChart.com.

72. Virilio 1989.

73. Cornebise 1984.

74. Kent 2001.

75. Turse 2008.

76. Der Derian 2001, 82.

77. The URL for this web page is http://ict.usc.edu.

78. Consalvo and Miller 2009.

79. Der Derian 2001, 166.

80. Ibid., 217.

81. Turse 2008.

82. Ibid., 270.

83. Stahl 2006, 125.

84. Ibid., 126.

85. Brown 2009, 2.

86. Ibid., 2.

87. Brown 2009.

WORKS CITED

Aarseth, E. 2001. Computer Game Studies: Year One. *Game Studies* 1. http://www.gamestudies.org/0101/editorial.html.

Acland, C. 2003. *Screen Traffic: Movies, Multiplexes, and Global Culture.* Durham, N.C.: Duke University Press.

Adorno, T., and M. Horkheimer. 2002. *Dialectic of Enlightenment.* Stanford, Calif.: Stanford University Press.

AFI's 100 Years . . . 100 Movies. 2006. *American Film Institute.* http://www.afi.com/tvevents/100years/movies.aspx.

Ahrens, F. 2003. Disney Board Loses Another Member. *Washington Post,* Dec. 2. http://web.ebscohost.com/ehost/detail?vid=3&hid=112&sid=60c5265b-59f4-48c0-bb80-0bb98b967564%40sessionmgr104&bdata=JnNpdGU9ZWhvc3QtbGl2ZQ%3d%3d#db=nfh&AN=WPT306060895503.

Ang, I. 1996. *Living Room Wars: Rethinking Media Audiences for a Postmodern World.* New York: Routledge.

Ault, S. 2007. Toshiba Says HD DVD Set-Top Players Are Back in Front. *Video Business,* Oct. 9. http://www.videobusiness.com/article/CA6488493.html.

———. 2008. Disc Spending Positively Flat at Midyear. *Video Business,* July 18. http://www.videobusiness.com/article/CA6579661.html.

Bart, P., and P. Guber. 2002. *Shoot Out: Surviving Fame and (Mis)Fortune in Hollywood.* New York: Perigee.

Barthes, R. 1977. *Image, Music, Text.* New York: Hill and Wang.

Block, A. 2008. ThinkFilm's Bergstein on Hollywood's Hot Seat. *Hollywood Reporter,* Aug. 5. http://www.hollywoodreporter.com/hr.content_display/news/e3ieb1f595b5fc21e0aa8a74a423ed2cb26?pn=1.

Bogost, I. 2007. *Persuasive Games: The Expressive Power of Videogames.* Cambridge, Mass.: MIT Press.

Brightman, J. 2007. Sony CEO: PS3 Momentum Now Similar to PS2. *GameDaily.* http://www.gamedaily.com/articles/news/sony-ceo-ps3-momentum-now-similar-to-ps2/18615.

Brookey, R. 2007. The Format Wars: Drawing the Battle Lines for the Next DVD. *Convergence* 13: 199–211.

Brookey, R., and P. Booth. 2006. Restricted Play: Synergy and the Limits of Interactivity in *The Lord of the Rings/Return of the King* Video Game. *Games and Culture* 1: 214–230.

Brookey, R., and R. Westerfelhaus. 2002. Hiding Homoeroticism in Plain View: The *Fight Club* DVD as Digital Closet. *Critical Studies in Media Communication* 19: 21–43.

Brown, S. 2009. Q&A: *Hobbit* Director Guillermo del Toro on the Future of Film. *Wired,* May 22. http://www.wired.com/entertainment/hollywood/magazine/17-06/mf_deltoro?currentPage=1.

Bryce, J., and J. Rutter. 2002. Spectacle of the Deathmatch: Character and Narrative in First-Person Shooters. In *Screenplay: Cinema/Video Game/Interfaces,* ed. G. King and T. Krzywinska, 66–80. London: Wallflower.

Burn, A., and D. Carr. 2006. Defining Game Genres. In *Computer Games: Text, Narrative and Play,* ed. D. Carr, D. Buckingham, A. Burns, and G. Schott, 14–29. Cambridge, U.K.: Polity.

Capodagli, B., and L. Jackson. 1999. *The Disney Way.* New York: McGraw-Hill.

Castronova, E. 2005. *Synthetic Worlds: The Business and Culture of Online Games.* Chicago: University of Chicago Press.

Castronova, E., J. Cummings, W. Emigh, M. Fatten, N. Mishler, T. Ross, and W. Ryan. 2009. Case Study: The Economics of Arden. *Critical Studies in Media Communication* 26: 165–179.

Catholic League for Religious and Civil Rights. 2000. http://www.catholicleague.com/disney/disney.htm.

Chown, J. 1988. *Hollywood Auteur.* New York: Praeger.

Christiansen, J. 2004. *Marvel Encyclopedia: Fantastic Four.* New York: Marvel Comics.

Company History. 2003. *Vivendi.* http://www.vivendi.com/corp/en/home.

Consalvo, M., and T. Miller. 2009. Reinvention through Amnesia. *Critical Studies in Media Communication* 26: 180–190.

Consumer Electronics Association. 2005. Historical Data. http://www.ce.org/Research/Sales_Stats/1219.asp.

———. 2006. *Five Technologies to Watch.* Arlington, Va.: Author.

Cornebise, A. 1984. *War as Advertised: The Four Minute Men and America's Crusade 1917–1918.* Philadelphia, Pa.: American Philosophical Society.

Corporate Information. 2007. *Sony Computer Entertainment Inc.,* Mar. 31. http://www.scei.co.jp/corporate/data/bizdataps_e.html.

Crawford, C. 2003. Interactive Storytelling. In *The Video Game Theory Reader,* ed. M. J. P. Wolf and B. Perron, 259–273. New York: Routledge.

Crowe, S. 2008. 87% of PS3 Owners Watch Blu-Ray on Console, Study Finds. *CEPro,* July 7. http://www.cepro.com/article/87_of_ps3_owners_watch_blu_ray_on_console_study_finds.

Daniels, B., D. Leedy, and S. Sills. 1998. *Movie Money: Understanding Hollywood's (Creative) Accounting Practices.* Los Angeles, Calif.: Silman-James.

Dekom, P. 2004. Movies, Money and Madness. In *The Movie Business Book*, ed. J. Squire, 100–116. New York: Fireside.
Del Toro to Direct *Hobbit*. 2008. *The Hobbit: The Official Movie Blog*, Apr. 24. http://www.thehobbitblog.com.
Der Derian, J. 2001. *Mapping the Military-Industrial-Media-Entertainment Network*. Boulder, Colo.: Westview.
Díaz, J. 2007. Unofficial Wii DVD Playback Available, Requires Modded Wii. *Gizmodo*, Aug. 10. http://gizmodo.com/gadgets/play-it-again,-mario/unofficial-wii-dvd-playback-now-available-requires-modded-wii-291060.php.
———. 2008. "Grand," but No "Godfather." *Wall Street Journal*, June 28. http://online.wsj.com/article/SB121460385251911957.html?mod=googlenews_wsj.
Dibbell, J. 2006. Owned! Intellectual Property in the Age of eBayers, Gold-Farmers, and Other Enemies of the Virtual State. In *The State of Play: Law, Games, and Virtual Worlds*, ed. J. Balkin and B. Noveck, 137–145. New York: New York University Press.
Documentary 1 for *The Fellowship of the Ring Extended Edition*. 2001. Peter Jackson, dir. New Line Cinema.
Dodd, A. 2005. *The Essential Guide to Telecommunications*. Upper Saddle River, N.J.: Prentice Hall.
Ducheneaut, N., N. Yee, E. Nickell, and R. Moore. 2006. Building an MMO with Mass Appeal. *Games and Culture* 1: 281–317.
Electronic Arts. 2004. Annual Report. *Electronic Arts Inc.* http://ccbn.mobular.net/ccbn/7/773/825.
Epstein, E. 2005. *The Big Picture: The New Logic of Money and Power in Hollywood*. New York: Random House.
Everett, A. 2003. Digitextuality and Click Theory. In *New Media: Theories and Practices of Digitextuality*, ed. A. Everett and J. Caldwell, 3–28. New York: Routledge.
Fahey, M. 2009. Gold Farming Website Sells for $10 Million. *Kotaku*, Jan. 29. http://kotaku.com/5141892/gold-farming-website-sells-for-10-million.
Filiciak, M. 2003. Hyperidentities: Postmodern Identity Patterns in Massively Multiplayer Online Role-Playing Games. In *The Video Game Theory Reader*, ed. M. J. P. Wolf and B. Perron, 87–102. New York: Routledge.
Fineman, J. 2005. Blockbuster Shares Soar. *Star-Telegram*, Oct. 5. http://www.dfw.com/mld/dfw/business/12846707.htm.
Finke, N. 2008. Paramount-Marvel Deal: Paramount to Distribute Next Five Films. *Nikki Finke's Deadline Hollywood Daily*, Sept. 29. http://www.deadlinehollywooddaily.com/exclusive-paramount-to-distribute-next-marvel-films.
Fiske, J. 1987. *Television Culture*. New York: Routledge.
Fritz, B. 2006. Taking a Whack: Stakes High for *Godfather* Video Game. *Variety*, Mar. 20. http://www.variety.com/article/VR1117940063.html?categoryid=13&cs=1.

———. 2008. Actors May Have a Shot at Videogame Residuals . . . in 2009. *Variety,* May 21. http://weblogs.variety.com/the_cut_scene/2008/05/actors-may-have.html.

GameStop Corp. 2008. *2007 Annual Report*. Delaware: Author. http://library.corporateair.net/library/13/130/130125/items/295079/2007annualreport.pdf

Ganapati, P. 2008. Blu-Ray Victory Won't Budge Microsoft. *TheStreet.com,* Feb. 21. http://www.thestreet.com/story/10404412/1/blu-ray-victory-wont-budge-microsoft.html.

Gantayat, A. 2006. *Dragon Quest IX* Q&A. *IGN Entertainment,* Dec. 12. http://ds.ign.com/articles/750/750610p1.html.

Garite, M. 2003. The Ideology of Interactivity (or, Video Games and the Taylorization of Leisure). Paper presented at the Level Up Games Conference, Utrecht, The Netherlands.

Gavin, A. 2003. Last Disney on Walt Disney Board Resigns with Angry Letter. *Orange County Register,* Dec. 1. http://web.ebscohost.com/ehost/detail?vid=9&hid=12&sid=60c5265b-59f4-48c0-bb80-0bb98b967564%40sessionmgr104&bdata=JnNpdGU9ZWhvc3QtbGl2ZQ%3d%3d#db=nfh&AN=2W60948578073.

Gershon, R., and T. Kanayama. 2002. The Sony Corporation: A Case Study in Transnational Media Management. *International Journal of Media Management* 4: 106–118.

Graser, M., and D. McNary. 2008. WB's Hero Hunt Heats Up. *Variety,* Aug. 15. http://www.variety.com/article/VR1117990663.html?categoryid=13&cs=1.

Green, E. 2009. Why the WGA Videogame Awards Are a Bad Thing. *Comic Book Bin,* Jan. 20. http://www.comicbookbin.com/wgavideogamewritingawardabadthing-001.html.

Grieb, M. 2002. Run Lara Run. In *Screenplay: Cinema/Video Game/Interfaces,* ed. G. King and T. Krzywinska, 157–170. London: Wallflower.

Griffin, N., and K. Masters. 1996. *Hit & Run.* New York: Touchstone.

Grimmelmann, J. 2006. Virtual Power Politics. In *The State of Play: Law, Games, and Virtual Worlds,* ed. J. Balkin and B. Noveck, 146–157. New York: New York University Press.

Grossberg, L. 1995. Cultural Studies vs. Political Economy: Is Anybody Else Bored with This Debate? *Critical Studies in Mass Communication* 12: 72–81.

Grover, R. 2007. *300*'s Lessons for Hollywood. *BusinessWeek,* Apr. 9. http://www.businessweek.com/bwdaily/dnflash/content/apr2007/db20070409_889613.htm?chan=top+news_top+news+index.

Hall, S. 1990. Cultural Identity and Diaspora. In *Identity: Community, Culture, Difference,* ed. J. Rutherford, 222–237. London: Lawrence and Wishart.

Handel, J. 2009. No More Sounds of Silence on the Music Composition Front. *Digital Media Law,* Nov. 18. http://digitalmedialaw.blogspot.com.

Hill, M. 2004. Despite Troubled CEO, Takeover Attempt, Disney Is Doing Just Fine. *Philadelphia Inquirer,* Feb. 29. http://web.ebscohost.com/

ehost/detail?vid=6&hid=108&sid=60c5265b-59f4-48c0-bb80-0bb98b96 7564%40sessionmgr104&bdata=JnNpdGU9ZWhvc3QtbGl2ZQ%3d%3d #db=nfh&AN=2W62088050255.

Hinley, P. 2008. Occupational Therapists Use Wii in Parkinson's Study. *MCG News,* Apr. 7. https://my.mcg.edu/portal/page/portal/News/archive/2008/Occupational%20therapists%20use%20Wii%20in%20 Parkinson's%20study.

Hocking, C. 2007. Hollywood's Bloody Ballet. *Click Nothing,* Apr. 21. http://clicknothing.typepad.com/click_nothing/2007/04/hollywoods_bloo.html.

Hollywood Movies Misfire with Core Audience. 2005. *Reuters,* Oct. 10. http://today.reuters.com/news/newsArticle.aspx?type=filmNews &storyID=2005-10-10T223207Z_01_DIT081053_RTRIDST_0_FILM -MEDIA-MOVIES-DC.XML.

Howells, S. 2002. Watching a Game, Playing a Movie: When Media Collide. In *Screenplay: Cinema/Video Game/Interfaces,* ed. G. King and T. Krzywinska, 110–121. London: Wallflower.

Hudson, K. 2004. Disney Shareholders Put Pressure on Company Board to Make Change at Top. *Denver Post,* Mar. 4, 1. http://web.ebscohost.com/ehost/detail?vid=6&hid=108&sid=60c5265b-59f4-48c0-bb80-0bb98b96 7564%40sessionmgr104&bdata=JnNpdGU9ZWhvc3QtbGl2ZQ%3d%3d #db=nfh&AN=2W60558868513.

Huizinga, J. 2000. *Homo Ludens: A Study of the Play-Element in Culture.* London: Routledge.

Hunter, W. 2000. Player 2 Stage 6: Laser Daze. *Dot Eaters,* June 10. http://www.thedoteaters.com/p2_stage6.php.

Interactive Contracts FAQ. 2009. *Screen Actors Guild.* http://www.sag.org/interactive-contracts-faq.

Investor Relations. 2009. *EA Games Company Information.* http://investor .ea.com.

Ivan, T. 2009. Analyst: No New Console Cycle before 2013. *Edge,* May 1. http://www.edge-online.com/news/analyst-no-new-console-cycle-before-2013.

Iwata, S. 2008. Message from the President. *Nintendo Annual Report 2008.* http://www.nintendo.co.jp/ir/pdf/2008/annual0803e.pdf.

Jenkins, D. 2007. Brash Entertainment to Debut as New Publisher. *Gamasutra,* Mar. 26. http://www.gamasutra.com/php-bin/news_index. php?story=13271.

Jenkins, H. 1992. *Textual Poachers.* New York: Routledge.

———. 2006. *Convergence Culture: Where Old and New Media Collide.* New York: New York University Press.

———. 2007. *Fans, Bloggers, and Gamers: Media Consumers in a Digital Age.* New York: New York University Press.

Johnson, D. 2007. Will the Real Wolverine Please Stand Up? Marvel's Mutation from Monthlies to Movies. In *Film and Comic Books,* ed. I. Gordan, M. Jancovich, and M. McAllister, 64–85. Jackson: University Press of Mississippi.

Keighley, G. 2007. Cameron Chooses Ubisoft for *Avatar* Game. *Hollywood Reporter,* July 25. http://www.hollywoodreporter.com/hr/content_display/news/e3ifbof8af7f8od3cacb8d27b004f9d7267.

Kent, S. 2001. *The Ultimate History of Video Games.* New York: Three Rivers.

Kerr, A. 2006. *The Business and Culture of Digital Games.* London: Sage.

Kim, W. C., and R. Mauborgne. 2004. Blue Ocean Strategy. *Harvard Business Review* 82: 76–84.

———. 2005. *Blue Ocean Strategy: How to Create Uncontested Market Space and Make the Competition Irrelevant.* Boston, Mass.: Harvard Business School Press.

Kinton, J. 2008. With Wii Bit of Help, Rehabbers Doing Fine. *Mansfield News Journal,* Dec. 27. http://www.mansfieldnewsjournal.com/article/20081227/NEWS01/812270307/1002.

Kline, S., N. Dyer-Witheford, and G. De Peuter. 2003. *Digital Play: The Interaction of Technology, Culture, and Marketing.* Montreal, Quebec: McGill-Queen's University Press.

Kohler, C. 2005. *Power Up: How Japanese Video Games Gave the World an Extra Life.* Indianapolis, Ind.: BradyGAMES.

Lang, D. 2008. *Dark Knight* Reigns Not in Video Game. *Associated Press,* Aug. 7. http://ap.google.com/article/ALeqM5gYbdYXq2-Geb9gVY4idWXCsh7Aa QD92DK9F80.

Lapsley, R., and M. Westlake. 1989. *Film Theory: An Introduction.* Manchester, U.K.: Manchester University Press.

Mactavish, A. 2002. Technological Pleasure: The Performance and Narrative of Technology in *Half-Life* and Other High-Tech Computer Games. In *Screenplay: Cinema/Video Game/Interfaces,* ed. G. King and T. Krzywinska, 33–49. London: Wallflower.

Marcus, P. 2002. *Spider-Man: Official Strategy Guide.* Indianapolis, Ind.: Brady.

Marshall, P. D. 2002. The New Intertextual Commodity. In *The New Media Book,* ed. D. Harries, 69–81. London: British Film Institute.

Marvel Entertainment, Inc. 2006. *2005 Annual Report.* New York: Author. http://www.marvel.com/company/pdfs/AnnualReport_March_24_2006_00032886.pdf.

———. 2008. *2007 Annual Report.* New York: Author. http://www.marvel.com/company/pdfs/10-k_as_filed_3-25_00054158.PDF.

Maslin, J. 1993. Review/Film: Plumbing a Video Game to Its Depths. *New York Times,* May 29. http://movies2.nytimes.com/mem/movies/review.html?_r=2&title1&title2=SUPER%20MARIO%20BROTHERS%20%28MOVIE%29&reviewer=Janet%20Maslin&v_id=47787&pdate=19930529&partner=Rotten%20Tomatoes&oref=slogin.

Mattelart, A. 1980. Cultural Imperialism, Mass Media and Class Struggle: An Interview with Armand Mattelart. *Insurgent Sociologist* 9: 69–79.

McAllister, M. 2001. Ownership Concentration in the U.S. Comic Book Industry. In *Comics and Ideology,* ed. M. McAllister, E. Sewell Jr., and I. Gordon, 15–38. New York: Peter Lang.

Member Agreement. 2009. *Disney's Pirates of the Caribbean Online.* http://disney.go.com/pirates/online/v3/help/memberagreement.html.
Mick, J. 2007. The First 9 Months: Blu-Ray Outsells HD DVD. *DailyTech,* Oct. 24. http://www.dailytech.com/The+First+9+Months+Bluray+Outsells+HDDVD/article.aspx?newsid=9390.
Miller, T. 2006. Gaming for Beginners. *Games and Culture* 1: 5–12.
Mitchell, E. 2003. Everyone's a Film Geek Now. *New York Times,* Aug. 17. http://www.nytimes.com/2003/08/17/movies/17ELVI.html?scp=1&sq=Everyone's%20a%20Film%20Geek%20Now%20Mitchell&st=cse.
Morley, D. 1980. *The Nationwide Audience: Structure and Decoding.* London: British Film Institute.
Movie Gallery Receives Final Approval of Debtor-in-Possession Financing. 2007. *CNNMoney.com,* Nov. 14. http://money.cnn.com/news/newsfeeds/articles/prnewswire/NYW20114112007-1.htm.
Nakamura, L. 2009. Don't Hate the Player, Hate the Game: The Racialization of Labor in *World of Warcraft. Critical Studies in Media Communication* 26: 128–144.
Nathan, J. 1999. *Sony: The Private Life.* Boston: Houghton Mifflin.
Ott, B., and C. Walter. 2000. Intertextuality: Interpretive Practice and Textual Strategy. *Critical Studies in Media Communication* 17: 429–446.
Ovadia, A. 2004. Consumer Products. In *The Movie Business Book,* ed. J. Squire, 447–456. New York: Fireside.
Perenson, M. 2008. Game Over: Format War Ends as Toshiba Drops HD DVD. *PC World,* Feb. 19. http://blogs.pcworld.com/staffblog/archives/006508.html.
Pixar, Inc. 2003. Pixar to End Talks with Disney. http://corporate.pixar.com/news/20040129-127764.cfm.
Poole, S. 2000. *Trigger Happy: Video Games and the Entertainment Revolution.* New York: Arcade.
Powers, A., and R. Brookey. 2005. Michael Eisner Was Framed: Newspaper Narrative of Corporate Conflict. In *Corporate Governance of Media Companies,* ed. R. Picard, 183–202. Jönkoping, Sweden: Jönkoping International Business School Research Reports.
Pustz, M. 1999. *Comic Book Culture: Fanboys and the True Believers.* Jackson: University Press of Mississippi.
Raugust, K. 1995. *The Licensing Business Handbook.* New York: EPM Communications.
Raviv, D. 2004. *Comic Wars: Marvel's Battle for Survival.* Sea Cliff, N.Y.: Heroes Books.
Raz, A. 1999. The Hybridization of Organizational Culture in Tokyo Disneyland. *Culture and Organization* 5: 235–264.
Rehak, B. 2003. Playing at Being: Psychoanalysis and the Avatar. In *The Video Game Theory Reader,* ed. M. J. P. Wolf and B. Perron, 103–127. New York: Routledge.
Reiner. 2006. *The Godfather:* The Game. *Game Informer* (May): 99.

Reisinger, D. 2008. Yep, Sony and Microsoft Sell Game Consoles too. *CNET News*, Dec. 26. http://news.cnet.com/8301-13772_3-10128950-52.html.
———. 2009. Why Video Game Developer Acquisitions Scare Me. *CNET News*, Mar. 5. http://news.cnet.com/8301-13506_3-10188950-17.html.
Remo, C. 2006. Activision Acquires Bond Video Game License. *Shacknews*, May 3. http://www.shacknews.com/onearticle.x/41980.
Rockstar in Hot Coffee over Mod. 2005. *Game Informer* (Sept.): 30.
Rose, M. 2004. Pixar to Disney: Sea You Later. *Hollywood Reporter*, Jan. 30. http://www.hollywoodreporter.com/thr/index.jsp.
Rosen, M. 1974. Frances Ford Coppola. *Film Comment* 4: 43–39.
Rosmarin, R. 2006. Nintendo's New Look. *Forbes*, Feb. 9. http://www.forbes.com/2006/02/07/Xbox-ps3-revolution-cx_rr_0207nintendo.html.
Rothman, W. 2004. DVD's? I Don't Rent. I Own. *New York Times*, Feb. 26, E1–6. http://www.nytimes.com/2004/02/26/technology/i-don-t-rent-i-own.html?scp=1&sq=rothman%20I%20don't%20rent%20I%20own%20DVD&st=cse
Rowles, D. 2006. Snakes Motherfucking Bites. *Pajiba*. http://www.pajiba.com/snakes-on-a-plane.htm.
Sarris, A. 1963. Notes on the Auteur Theory in 1962. *Film Culture* 28: 1–51.
Satariano, A. 2009. Activision Will Consider Acquisitions amid Recession. *Bloomberg.com*, Mar. 4. http://www.bloomberg.com/apps/news?pid=newsarchive&sid=aHj2TowkztkU.
Sawyer, B., A. Dunne, and T. Berg. 1998. *Game Developer's Marketplace*. Albany, N.Y.: Coriolis Group.
Schiller, H. 1979. Transnational Media and National Development. In *National Sovereignty and International Communication*, ed. K. Nordenstreng and H. Schiller, 21–32. Norwood, N.J.: Ablex.
Schumacher, M. 1999. *Francis Ford Coppola: A Filmmaker's Life*. New York: Crown.
Scott, A. O. 2007. Battle of Manly Men: Blood Bath with a Message. *New York Times*, Mar. 9. http://www.nytimes.com/2007/03/09/movies/09thre.html?ex=1181102400&en=84471177fd2ec9e6&ei=5070.
Sedman, D. 1998. Market Parameters, Marketing Hype, and Technical Standards: The Introduction of the DVD. *Journal of Media Economics* 11: 49–58.
Sheff, D. 1993. *Game Over: How Nintendo Zapped an American Industry, Captured Your Dollars, and Enslaved Your Children*. New York: Random House.
Sherry, J. 2001. The Effects of Violent Video Games on Aggression: A Meta-Analysis. *Human Communication Research* 27: 409–431.
Sinclair, J. 2004. Globalization, Supranational Institutions, and Media. In *The SAGE Handbook of Media Studies*, ed. J. Downing, D. McQuail, P. Schlesinger, and E. Wartella, 65–82. Thousand Oaks, Calif.: Sage.
Slater, M., K. Henry, R. Swaim, and L. Anderson. 2003. Violent Media Content and Aggressiveness in Adolescents: A Downward Spiral Model. *Communication Research* 30: 713–736.

Smith, D. 2005. The Video Game with an Offer You Can't Refuse. *Guardian Unlimited,* Apr. 17. http://observer.guardian.co.uk/international/story/0,6903,1461704,00.html.
Smith, S., K. Lachlan, and R. Tamborini. 2003. Popular Video Games: Quantifying the Presentation of Violence and Its Context. *Journal of Broadcasting and Electronic Media* 47: 58–76.
Snider, M. 2003. On DVD Menu: Extra Scenes, Extra Bucks for Studios. *USA Today,* Feb. 17. http://web.ebscohost.com/ehost/detail?vid=17&hid=114&sid=e582d05c-517a-4844-8d8e-ef4cabf7c49f%40sessionmgr101.
Stahl, R. 2006. Have You Played the War on Terror? *Critical Studies in Media Communication* 23: 112–130.
Surette, T. 2006. EA Settled OT Dispute, Disgruntled "Spouse" Outed. *GameSpot News,* Apr. 26. http://www.gamespot.com/news/6148369.html.
Taylor, T. 2006. *Play between Worlds: Exploring Online Game Culture.* Cambridge, Mass.: MIT Press.
Thorsen, T. 2005. Steven Spielberg, EA Ink Three-Game Next-Gen Deal. *GameSpot News,* Oct. 14. http://www.gamespot.com/news/6135746.html.
———. 2006. EA Sees $16 Million Quarterly Loss, Annual Revenue Decline. *GameSpot News,* May 3. http://www.gamespot.com/news/6148917.html?tag=result;title;0.
———. 2007. Sony Game Unit's Q2 Losses Double. *GameSpot News,* Oct. 25. *http://www.gamespot.com/news/6181712.html.*
Tong, W. L., and M. C. C. Tan. 2002. Vision and Virtuality: The Construction of Narrative Space in Film and Computer Games. In *Screenplay: Cinema/Video Game/Interfaces,* ed. G. King and T. Krzywinska, 98–109. London: Wallflower.
Turow, J. 1997. *Media Systems in Society.* New York: Longman.
Turse, N. 2008. *The Complex: How the Military Invades Our Everyday Lives.* New York: Metropolitan.
U.S. Residential Broadband Penetration to Exceed 50% in 2007. 2007. *Parks Associates,* Feb. 15. http://www.parksassociates.com/press/press_releases/2007/dig_lifestyles1.html.
Van Maanen, J. 1992. Displacing Disney: Some Notes on the Flow of Culture. *Qualitative Sociology* 15: 5–35.
Virilio, P. 1989. *War and Cinema: The Logistics of Perception,* trans. Patrick Camiller. London: Verso.
Vital Stats. 2007. *GameSpot.* http://www.gamespot.com/ps2/action/thegodfather/index.html.
Wexman, V. W. 2003. Introduction. In *Film and Authorship,* ed. V. W. Wexman, 1–18. New Brunswick, N.J.: Rutgers University Press.
White, M. 2008. Sony's PlayStation 3 Gains Ground on Xbox with Games. *Bloomberg.com,* July 14. http://www.bloomberg.com/apps/news?pid=20601080&sid=aClqhxdVor6g&refer=asia.
Wii-habilitation: Using Video Games to Heal Burns. 2008. *Weill Cornell News: Science Briefs,* June–July. http://www.med.cornell.edu/science/

sci/new/science-briefs-junejuly-2.shtml. Williams, D. 2002. Structure and Competition in the U.S. Home Video Game Industry. *International Journal on Media Management* 4: 41–54.

Williams, D. 2002. A Structural Analysis of Market Competition in the U.S. Home Video Game Industry. *International Journal on Media Management*, 4(1); 41–54.

Wolf, M. J. 1999. *The Entertainment Economy: How Mega-Media Forces Are Transforming Our Lives.* New York: Times Books.

Wolf, M. J. P., ed. 2001. *The Medium of the Video Game.* Austin: University of Texas Press.

———. 2003. Abstraction in Video Games. In *The Video Game Theory Reader,* ed. M. J. P. Wolf and B. Perron, 47–65. New York: Routledge.

Woodcock, B. 2008. MMOG Subscriptions Market Share—April 2008. *MMOGChart.com.* http://www.mmogchart.com/Chart7.html.

World of Warcraft Subscriber Base Reaches 11.5 Million Worldwide. 2008. Blizzard Entertainment, Dec. 23. http://www.blizzard.com/us/press/081121.html.

Writers Guild of America and Marvel Studios Announce Interim Agreement. 2008. Writers Guild of America, East, Jan. 25. http://www.wgaeast.org/index.php?id=285&tx_ttnews[tt_news]=1339&tx_ttnews[backPid]=-1&cHash=9ecb33f072.

Zacharek, S. 2007. *300. Salon,* Mar. 9. http://www.salon.com/ent/movies/review/2007/03/09/300.

Zimmerman, E. 2004. Narrative, Interactivity, Play, and Games: Four Naughty Concepts in Need of Discipline. In *First Person: New Media as Story, Performance, and Game,* ed. N. Wardrip-Fruin and P. Harrigan, 154–163. Cambridge, Mass.: MIT Press.

INDEX

Robert Alan Brookey is Associate Professor of Communication Studies at Northern Illinois University and author of *Reinventing the Male Homosexual* (Indiana University Press, 2002).